Tourist Destinations According to Stakeholder Strategies

Tourism and Mobility Systems Set

coordinated by
Philippe Violier

Volume 7

Tourist Destinations According to Stakeholder Strategies

From Clusters to Local Tourist Systems

Pierre Louart
Jérôme Piriou
Philippe Violier

iSTE

WILEY

First published 2024 in Great Britain and the United States by ISTE Ltd and John Wiley & Sons, Inc.

ISTE Ltd
27-37 St George's Road
London SW19 4EU
UK

www.iste.co.uk

John Wiley & Sons, Inc.
111 River Street
Hoboken, NJ 07030
USA

www.wiley.com

Library of Congress Control Number: 2024934847

British Library Cataloguing-in-Publication Data
A CIP record for this book is available from the British Library
ISBN 978-1-78630-659-3

Contents

Introduction

Tourist Destinations: At the Crossroads between Geographic and Organizational Proximity

I.1. The origins of this reflection

Reflections on the challenging relationships between geographic and organizational proximity are nothing new in spatial economics. The initiative can be attributed to Alfred Marshall. Actually, the interaction between the strategies of economic actors and decision-making in the capitalist world leads to a duality between, on the one hand, the places that concentrate companies under the aegis of agglomeration economies (even though impaired by certain limitations); and, on the other hand, marginalized territories. Economic dynamics are influenced by their demographic counterpart to such an extent that desertification threatens depreciated areas. There are certainly some exceptions: entrepreneurs who decide to settle in their homeland despite all odds, insofar as radical innovation can be a game changer and cause the abandonment of formerly privileged spaces. The mining basins in north-western Europe offer good examples of these reversals of fortune. Spatial concentration often reflects a concentration of companies, encouraged by the search for economies of scale.

By the end of the 19th century (1890), Alfred Marshall noted that between metropolises and the margins, certain spaces flourished, which – far from being defined as residual forms or by the use of outdated techniques (Gaffard and Romani 1990) – could be considered as innovative environments, insofar as non-market relationships and certain forms of quasi-integration enabled them to develop and remain at the cutting edge of progress. This reflection was later pursued, particularly in Italy. Giacomo Becattini relied on it to explain the dynamism of "The Third Italy", located in the center of the peninsula and which, after World War II, offered a growth model based on original principles, which differed from both the Mezzogiorno and the classic mechanisms at work in the northern region.

The analysis was then taken up by numerous authors, including Michael Porter, who focused on the concept of *cluster*, now very popular in Quebec.

Porter (1998, p. 197) provided the following definition:

> A cluster is the geographical concentration of interconnected companies: suppliers and service providers in related industries and associated institutions; the firms delivering the final product cooperate with universities, as well as with their competitors.

In 2000, Porter emphasized the importance of social relations between more or less formal actors and refined his approach by assimilating the cluster to a territorially integrated value chain. Even though the terms are often used synonymously, clusters differ from simple networks (Bédé 2015). For Thorelli (1986), a network comprises two or more companies, linked by exchange relationships which last over time, and whose stability makes room for reciprocal trust between the actors. However, as Tinsley and Lynch (2001) pointed out, the network is not exclusively dependent on space. In other words, geographical proximity is not a sufficient condition for it.

I.2. The fortune and misfortune of words

Although other expressions (such as Localized Industrial System) are used, the term "cluster" seems to have prevailed. Cluster is now such a popular term that it has ended up designating realities that are unrelated (or poorly related) to the definition provided by Porter. For example, Atout France set up thematic clubs for stimulating exchanges among professionals identified as coming from different backgrounds. Therefore, there was a coastal club, an urban club, etc. Then, the term "club" was replaced by that of "cluster". However, in Porter's mind, the concept of cluster combines organizational proximity (quasi-integration) with geographical proximity (e.g. a small distance). Describing a national-scale network as a cluster amounts to considering that distance is irrelevant. However, it is an essential explanation not only of the agreement, trust and cooperation construction between companies, but also of a certain dynamism and a ferment of ideas leading to innovation. In this sense, the concept of cluster involves an innovative environment, in the spirit of GREMI (Tabariés 2005), emphasizing the fact that inter-organizational relations can be influenced by the territory.

The term cluster continues to be increasingly used. During the Covid-19 pandemic, the term was used for denoting particularly active outbreaks of infection. Since then, it has lost much of its relevance.

At some point, the term was replaced by that of ecosystem, imported from life sciences and adapted by Moore (1996). Although Glon and Pecqueur (2016) explained that territorial ecosystems are sets of private and public actors establishing relationships with one another and with their territory, and who pursue collaborative goals beneficial to all, the use of this term raises certain questions when we trace back its original meaning. In point of fact, the ecosystem describes the relationships that exist between living beings and their environment, within a biotope. It therefore describes mechanical and compulsory links. Transferring this idea to the social sciences amounts to minimizing the still dominant part of the actors' strategies.

For this reason, the authors of this book have chosen to activate an already old expression that has retained its full meaning: Localized Industrial Systems, and in the case of tourism, Localized Tourism Systems (LTS).

I.3. Aim of the book

The aim of this work is to analyze the particular conditions in which such systems are deployed in the field of tourism. In this work, tourism will be understood as a system whose purpose is the recreation of individuals (Knafou and Stock 2013[1]). Other forms of motility will be intentionally left aside (despite institutions and a portion of researchers persisting in the idea of integrating them), considering that the differences in nature are such that an amalgamation does not enable us to understand what is at stake in these movements, based on individual projects and on the use of completely free time. As Elias and Dunning (1995) pointed out, "traveling while on vacation" constitutes a specific moment and an act that makes it possible to combat routine actions most effectively:

> By "routine actions" we mean recurrent channels of action enforced by interdependence with others, and which impose upon the individual a fairly high degree of regularity, steadiness and emotional control in conduct and which block other channels of action even if they correspond better to the mood, the feelings, the emotional needs of the moment (Elias and Dunning 1995, p. 80).

1 A first edition of this dictionary included the entry "tourism" in 2003. The new edition has not made any changes to this text.

Even though tourism was the heart of the investigation, other worlds were also studied as soon as a pattern could be found. This was the case, for example, of the production of spirits or culture.

Our approach[2] began with case studies, chosen on the grounds of their relevance. These analyses are detailed in Chapter 1. This is followed by two reflective chapters. The first of these is tackled from the perspective of two geographers. The second one translates the views of two specialists in management sciences. These disciplines differ in their relationship to action. While social geography is concerned with analyzing the strategies of actors placed in a particular local context, managers are keen on investigating the actions put in practice by individuals and their organizations to carry out their projects, and on issuing recommendations. The authors considered that such complementarity (which makes a multidisciplinary approach relevant) should be reflected in the book's construction. Each perspective is presented in a different chapter. Chapter 4 proposes a typology of clusters applied to the field of tourism and develops recommendations for action.

2 This work was built as a continuation of the research program "CONNECT: *Construction, knowledge, evolution of the* tourism *cluster*. Innovation in destinations' governance and development of tourism clusters" financed between 2016 and 2020 by the Pays de la Loire region, via the *Recherche innovation formation* (RFI), Angers Tourism Lab.

1

Case Studies

The analyses discussed in this work are based on case studies. They will be exposed in this first chapter, presented one after the other, with no pre-established order other than alphabetical.

1.1. Amnéville or the extravagant invention

Amnéville-les-Bains unexpectedly emerged in the middle of the industrial wastelands. Why and how did this happen? And what became of this destination after the death of its creator?

1.1.1. *The opportunistic idea of an entrepreneurial mayor*

The transformation of Amnéville into an innovative tourist area was the result of Jean Kiffer's work. This doctor was the city's mayor from March 1965 to August 11, 2011, for eight successive terms, the last of which was interrupted by his death. He was also a deputy for three terms: member of the RPR (Rally for the Republic), and then the DVD (*Divers Droite*).

After the crisis in the Lorraine steel industry, Jean Kiffer transformed industrial wastelands into tourism and leisure-dedicated areas, obtained a spa resort label in 1987, which enabled him to open a casino. By the time he died, the tourist site had generated 2,000 direct jobs.

Confronted with the decline of local industries, Jean Kiffer became an active promoter of lucrative leisure activities and contented himself with bringing jobs and economic stability to the city's inhabitants.

For a color version of all of the figures in this chapter, see www.iste.co.uk/louart/touristdestinations.zip.

Until the 2000s, Jean Kiffer benefited from a period during which the authorities were not too meticulous as to the methods used, and where certain regulatory or administrative privileges were accepted in exchange for positive socio-economic effects. From an industrial point of view, the region was in so much decline that, despite sounding too adventurous, the options encouraged at the moment ended up proving profitable, as they contributed to the economic momentum and generated profits.

A century later, Jean Kiffer could still illustrate Fayol's theories on the business manager (Hatchuel 2019). Contrary to what has been said in reductionist analyses, Fayol did not aim to exalt the manager's "temperament", nor to highlight the art of giving orders or being obeyed.

> He assigned himself complex and surprising missions for his time, such as stimulating a permanent and indefinite improvement of the activities he was responsible for, and acting as a political leader (*sic*), attentive to giving substance to the social body of his company (Hatchuel 2019, p. 95).

In this perspective, Fayol strove to combine "the general and the private interest" (ibid., p. 96). To be engaging, he emphasized the need for a "general action program", which could be the subject of directives when the future was predictable, but could become "an adventure, when the unknown was major" (ibid., p. 102). Finally, when Fayol spoke of foresight, he implied "protecting oneself against known dangers and guaranteeing the regeneration of forces for unknown future battles", which required "mutual aid, solidarity and social justice" (ibid., p. 103).

In the case of Kiffer, the tyranny of an authoritarian and proactive project manager was reinforced by the pressing need to regenerate the territory. Confronted with unimaginative bureaucrats, he took pleasure in showing that we could actually change the context and invent a different type of economy, based on tourism and leisure. He transformed the a priori depressing image of a city deemed as devastated. In doing so, he relaunched works and modified the city's image to mirror his own views. He even built a second city, juxtaposed to the first, over former mining sites.

In order to achieve the results described hereafter, Jean Kiffer managed to elicit the residents' trust (who made the most from financial benefits) and national support (through his political affiliation, relationships and accumulated experience as an elected official). He acted in phases: a gradual start, followed by a massive deployment when greater resources arrived, and a continuation of activities against all odds – even when difficulties began to accumulate – until his sudden death in 2011.

Jean Kiffer designed his leisure city as a global product obeying a traditional management life cycle (preliminary tests, then a leap forward and seizing favorable circumstances, and finally the management of a maturity phase, followed by the renewal of needs in the face of a more difficult context or new competition).

At the economic level, Amnéville was built as a conglomeration of activities[1], accompanied by the strengths and weaknesses of this type of model. On the strengths side, this model makes it possible to draw in an audience with diverse needs, means or interests, who can choose from a variety of attractions. As far as weaknesses are concerned, it is difficult to excel in every aspect. Insofar as quality facilities compete with those in bigger cities (e.g. in Metz, Luxembourg, at Center Parcs), there may be a leak in visitor engagement, except among those who prefer the proximity of "everything on spot". Besides, in a more demanding and strictly controlled economic system, business models may prove less profitable.

However, let us focus once again on Kiffer's initial intuition, fostering leisure and tourism to reinvent the local economy. At a deeper level, he was one of the pioneers in territorial decolonization (at a time when Parisian centralism was even more significant than nowadays). Throughout his life, he strove to reveal the innovation potential of territories, as well as their ability to become independent.

By the time of Kiffer's death, Amnéville had 10,000 inhabitants (compared to 8,000 during his first mandate), which translates the unforeseen development of a town having experienced the end of the steel industry.

1.1.2. *The rise of a popular park with multiple attractions*

After the German annexation of 1871, and until 1918, Amnéville was known as Stahlheim (the city of steel). Workers' towns were built near steel factories. At the beginning of the 20th century, the town had less than 5,000 inhabitants. It should have experienced its deindustrialization like the rest of Lorraine. From the moment he was appointed (1965), Doctor Kiffer began developing common public facilities (for sports, youth and the elderly). It was during his second mandate (from 1971) that he decided to transform the landscape into a City of Leisure, by combining public facilities and private creations, to match the opportunities that arose. In the 1980s, growth was rapid, Kiffer was in a hurry. As soon as revenue was generated, he invested in new projects.

From the outset, Kiffer acted as a visionary and created large facilities to attract a large audience. Since attendance was still low at the time, he could satisfy the demand and make the place more enticing.

---

1 This is not the Puy du Fou model, also deployed from a centralizing pilot, but primarily based on the same expertise shown elsewhere (the art of putting myths and history fragments on display). The reader may refer to the 2021 special issue of *Point* on Puy du Fou ("a culture of challenge and commitment"), coordinated by Catherine Golliau.

The acceleration of the trend began with the exploitation (in 1986), of ferruginous water at 41°C, known for treating rheumatologic and respiratory conditions. This made it possible to invent the concept of "Amnéville-les-Thermes" and, three years later, to install a casino, which in turn became a source of income and investment for new activities.

Keeping his composure, Kiffer aimed to turn Amnéville into the leading thermal center in France (by democratizing access to this type of care). Besides the spa center, mainly attended by rather elderly people (Saint Eloy, 1986), a fitness area for all audiences was created (Thermapolis, 1996); and, seven years later (2003), a luxurious establishment for wealthier clients (Villa Pompeii).

Figure 1.1. *Villa Pompeii, the last balneotherapy establishment built in Amnéville in 2013. Picture by Jérôme Piriou, 2019*

In the meantime, a zoo was established in the neighboring town of Hagondange (1986), but was called "Amnéville Zoo", which helped to broaden the site's potential centers of interest. In 1987, a tourist office (*syndicat d'initiative*) was opened.

Now, let us discuss the issue of the casino (currently owned by the Georges Tranchant Group). Its installation (1989) was made possible thanks to the existence of the thermal baths. In 1988, Charles Pasqua (at the time, Minister of the Interior and an acquaintance of Kiffer) issued an authorization to equip casinos with slot machines ("one-armed bandits"). This made them accessible to a popular clientele and made them highly profitable[2], by providing considerable *royalties* to the cities having them on their territory[3]. Amnéville benefited from an abundant source of income, which made it possible to invest in new projects (double extension of the thermal baths, *Galaxie* performance hall, cinemas, museums, sports activities, and an indoor ski slope).

As to the scope of the supply, the facilities are large everywhere.

The thermal facilities (Saint-Eloy, Thermapolis, Villa Pompeii) occupy three sites with varying functions and which cater to different audiences, accommodating not only spa guests, but also users seeking relaxation and curious day-visitors. In 2017, there were over 600,000 visitors, including nearly 16,000 subsidized curists for an 18-day period (8th rank among 110 French thermal establishments).

The Seven Casino has 350 slot machines and gaming tables, the fourth largest in France.

There are numerous sports facilities (multi-activity park, lake, golf course, swimming pool, ice rink, and treetop adventure course). The Snowhall is the only indoor ski slope in France (200,000 visitors annually, or a quarter of the skiers in the Vosges massif).

The *Galaxie* performance hall has 12,000 seats, hosting important stars and shows. There is also a multiplex cinema, managed by Gaumont, with 12 screens.

In total, the activity list spans 70 possibilities (from miniature golf to the aquarium, from summer tobogganing to pedal boats on the lake). The promotional brochure claims real pride in having created a tourist resort in the middle of nature, "over former steel slag heaps".

It is nonetheless difficult to determine what the 6 million visitors announced annually correspond to (most of them being counted on several facilities, or several

---

2 "Representing 83% of the gross gaming revenue, the 10,684 slot machines installed exclusively in 137 of the 154 French casinos transformed this type of leisure in seven years. When their operation was authorized, they generated 1 billion francs in earnings every year for traditional games alone, a panorama which had been declining since the mid-1970s. However, slot machines (actual piggy banks programmed to retain up to 15% of the sums played), have made casinos earn no less than 1 billion additional francs every year […]. Such a financial upheaval did not occur without a profound change in the game culture". Formerly reserved for high society, "casinos now homogeneously welcome all socio-professional categories", Mr. Castaing, *Le Monde*, February 13, 1990.

3 Casinos must return a significant portion of their profits to host communities.

activities). The length of stay on the site is also unknown, as there are several types of accommodation (approximately 15 restaurants, 13 hotels, *bungalow* rental units, etc.).

Behind the concept of Amnéville, there is the desire to offer sensations, the pleasure or relaxation (health restoration), at a reasonable price, for a population with limited resources. At the same time, visitors have the ease of trying a bit of everything and, if necessary, making their choice on site, depending on the moment's desires.

Figure 1.2. *Leisure activities within walking distance. Picture by Jérôme Piriou, 2019*

We may ski one morning, see the zoo in the afternoon and, why not, spend the following day at the thermal baths. Or, within the same family, each member may follow their distinct pleasure. While it is true that we may jump from one place to another, this option is not cheap since everything is immense and spectacular. Personal or seasonal events can also be organized (family gatherings and cousin fests, Valentine's Day, Halloween, etc.).

Admittedly, there is a bit of patchwork, and everything is not clean or perfect. However, who could say such about a city, even a prestigious one? In Germany, Europa Park, one of the most visited parks in Europe, also offers atmospheres of all kinds, which are both evocative and kitsch, with sporting options for all tastes, and an abundance of places to stay, eat or drink.

1.1.3. *Excesses or shortages with serious long-term consequences*

Dr. Kiffer was authoritarian. He was driven by a healthy kind of rage, trying to make Amnéville escape its reputation as an ominous town ("Longwy perhaps, but not us"). Despite his excesses, he had a great capacity for recovery, which earned him the support of the city's inhabitants. After all, he had good ideas, projects, an ability to publicize the city's attractiveness and to invent opportunities for people to come or stay there.

As C. Rollot has shown[4], thermal opportunities span a wide variety, from the ordinary cure to the aquatic circuits, from light therapy to the fitness space, from cryotherapy to massage mattresses. Most importantly, the mayor welcomed the new residents, and helped them to get settled and find jobs. By 2018, "eighty percent of the four hundred and sixty-five employees at the thermal establishments came from Amnéville or the neighboring communities"[5].

However, the mayor's imperious grip – until then an apparent source of speed and efficiency – brought about three disadvantages. Although he was able to promptly sweep them away during the period of strong growth, their effects would be felt afterwards, since the beginning of the 2000s; and even more so, after his death in 2011.

1.1.3.1. *The headlong rush (financial risks and the undermining of bureaucratic constraints)*

Jean Kiffer's art came down to acting quickly, by associating his establishments with an alluring commercial logic and an innovative concept of popular tourism. As shown in a newspaper cutting from the 1990s[6], he went to great lengths to find new customers.

---

4 C. Rollot, *Le Monde*, July 5, 2018, *Amnéville-les-Thermes, du bassin sidérurgique aux bassins aquatiques*.

5 C. Rollot, *Le Monde*, July 5, 2018.

6 "Actually, everything happened very quickly after the detection of a 'miraculous spring', Saint-Eloy, which had been flowing there since the dawn of time and whose virtues no one really knew, before the experts revealed its properties for treating rheumatism and certain respiratory conditions. The water table was found at a depth of 900 meters (first drilling in 1979) and the thermal center could open its doors in June 1986 […]. Its success was immediate at the local level. Amnéville established itself as the first spa resort in Lorraine, ahead of, in order, famous spa towns as Vittel, Plombières-les-Bains and Contrexéville. Doctor Kiffer had the genius idea of creating 'a local spa' by setting up minibus shuttles, up to 50 kilometers around – that is to say, reaching Luxembourg and West German borders – in a region where silicosis is widespread. At the same time, three hotel-restaurants were built, including one by former steelworkers who invested their severance pay and a block of furnished studios and apartments to accommodate the elderly, in particular". Not only did visitors come for treatment, but also to relax. "Alongside sports facilities, leisure places of all kinds sprouted like mushrooms".

Meeting no real opposition at home (there was no opposition list in his last municipal elections), Kiffer always claimed to have overcome challenges (making the most from opportunities) and often amused himself by "confronting the administration with the fait accompli". According to his closest acquaintances, "he embodied the local law". He was not afraid of offending his colleagues from neighboring towns, having an extensive and flexible conception of the territory. The advantage, for him, was to move at full speed, even though there was a risk of a headlong rush.

To move forward, Kiffer used professional taxes: while Unimétal, the first French factory for long products (cast iron sector), paid a little over 20 million francs in taxes per year, in 1989, the casino became the second largest taxpayer in Amnéville, paying nearly 10 million francs annually. In Kiffer's words, this made room for a kind of "municipal capitalism". In this spirit, with far-fetched dreams, he acted by mixing public resources with paternalistic private management.

This led to a meta-organization[7] logic, where politics became involved in all aspects, playing on the snowball effect of the process. Commercial success brought in money and optimized accounts, especially thanks to abundant tax revenues. In this conglomeration of opportunities, there was no need to excel in all aspects, because competition was still weak and the appeal to popular audiences was innovative. It was a good illustration, before its time, of the Blue Ocean Strategy[8] (according to which we should compete in blue oceans, devoid of intense competition).

In this situation, the mayor played a key role and kept a tight grip on every issue surrounding him, playing on several tables at the same time. Opportunistic, economically shrewd, unconventional, Kiffer used and abused his territorial sovereignty, until his borderline "arrangements" were finally called into question. For him, obeying stringent laws came down to breaking the dynamism he had been able to establish.

Kiffer's point of view could be understood. Many public rules are restrictive and in conflict with the creative opportunities of entrepreneurship, which require acting quickly or confronting competition. Regulation may undermine creation ("in France,

7 This notion has been well conceptualized by various authors in organization theory, in particular by A. Goran and N. Brunsson. A summary of their analysis was produced by H. Dumez (*Annales des mines, Gérer et comprendre*, 2009, 95, p. 77–78). The reader may also refer to an article by V. Leys and P. Joffre, *Méta-organisations et évolution des pratiques managériales*, *RFG*, 214–4(241), 121–134.
8 W. Cham Kim, R. Mauborgne, 2009.

the administrative time for creating or managing economic activities doubles that of Germany, and triples that of the United Kingdom"[9]).

However, in the case of Kiffer, this resulted in a sort of administrative cavalry, camouflaging the fact that certain operations had bypassed the legal framework and could lead to administrative, commercial or even criminal disputes. He ignored respect for certain rules, hoping for a logical fait accompli, given the difficulty of turning back the clock.

Over time, the mayors from surrounding districts gained further power, the administration woke up, and greater demands challenged cookie-cutter solutions. The requests for adjustments followed one another.

Kiffer was condemned on several occasions by the courts and indicted by the Court of Auditors for his management of Amnéville. As his anger increased, shortly before his death, he wanted to rename the city by its German name, declaring it "the virtual Principality of Stahlheim" (resistant as steel). "The Republic, which has only harmed us, the Republic that we never needed. The Republic, I leave it aside"[10].

At the same time, territorial benchmarks were changing. The relevant environment was expanding, Metz had awakened (thanks to the Pompidou Center inaugurated in 2010), Luxembourg was developing its competition and parallel attractions were being organized, in the same region, for other audiences.

1.1.3.2. *The nepotistic management of activities*

Jean Kiffer, a centralizer and a despot, surrounded himself with loyal people and whom he promoted in the structures he created or developed, except in the autonomous establishments which he included in his territorial space (namely, the zoo and the casino). He always had a preference for commitment (affiliation) over technicality or managerial competence, judging that he could manage with the necessary expertise on his own.

This nepotistic distribution of responsibilities had two disadvantages. While it is true that the employees who owed their position to Kiffer (due to family, emotional or political reasons) were grateful to him, they kept their jobs regardless of their actual results, without necessarily seeking operational improvement or financial optimization. They did not always have the required managerial qualities. Furthermore, they were "merely following orders" and had little means to challenge certain hierarchical choices, even in case of objection.

9 Comments made in 2018 by a manager who owns companies in the three countries.

10 This was the theme of his last greeting speech to the population of Amnéville, in January 2011.

1.1.3.3. *The creation of an amalgam, with juxtapositions lacking synergy*

Overall, the tourist town of Amnéville had become a sort of amalgam (in the sense of an alloy, a mixture, a combination of opportunities). While in some respects it worked properly – by creating a set of mixed interactions – in others, it gave the impression of being a heterogeneous and mismatched system, made up of bits and pieces.

This is not necessarily a problem for a tourist destination, since the same place can attract various kinds of people, tackling different itineraries, provided that these surroundings do not repel one another. In this case, the mix was rather favorable, since the supply was aimed at rather popular audiences, even though Amnéville offered a range of more selective activities.

Even though the tourist supply seemed insufficiently coordinated – due to its gradual expansion following opportunistic developments – it adapted quite well to contradictory or complex tourist expectations (depending on the people or groups, as well as at the individual level).

After all, amalgamation is the business model of city centers. It must be varied, there must be something appealing to all tastes, but with a minimum of homogeneity in relation to criteria such as the people's *habitus*, their social marking, etc. There may be preliminary neighborhood choices, although this is debatable.

However, there are two potential drawbacks which should have been taken into account:

– The absence of real synergy: by juxtaposing attractions, we do not necessarily optimize their complementarity.

– The lack of quality in part of the supply: in a conglomerate logic, we may lack expertise, or even leave some activities unsupervised, with any problems arising in terms of production or marketing. For example, this is what came upon *Galaxie* (poorly chosen schedule) or the zoological park (erratic management), with further repulsive effects for the audience (poor equipment maintenance, animal mistreatment, non-respect of their remains, expensive shows without a good return on investment). We also risk losing customers to competition whose supply is more technical or more qualified.

This quality competition appeared after the beginning of the 2000s.

For example, the supply at the thermal area of Mondorf-les-Bains was enriched (set on the France–Luxembourg border, just 36 km and 35 minutes away). It is the only spa center in Luxembourg, with a stylish park filled with works of art, art nouveau houses and a casino. The city of Metz strengthened its cultural policy through the development of its downtown area, the enhancement of its cathedral and the relocation of the Beaubourg National Center for Art and Culture; also, a neighboring Center Parcs is almost physically attached to the city. Although there is a factual market segmentation, Amnéville, Metz and Mondorf-les-Bains do not cater to the same audiences. Then, the question arises: Can a leisure city be exclusively in the low end of the supply? The creation of Villa Pompeii revealed an opposite desire.

1.1.4. *Three challenges for the future: cleaning up the accounts, expanding audiences, optimizing the supply*

When the creator of Amnéville City of Leisure died, the new scenario was challenging: increased competition, a conglomeration of establishments difficult to manage, tactical errors on the part of certain actors, financial problems, increased administrative requirements, and less invested appointed officials who sought to protect themselves rather than to act on the margins of legality.

The visitors were still there, their numbers satisfactory, but behind the scenes was a different reality.

Succeeding Kiffer as mayor, Doris Belloni (former deputy mayor) was not easily elected at the end of 2011. Everyone knew there were adjustments to be made. She promised an audit of municipal finances and pledged to "protect the city to the best interest of its residents". It was evident she tried to free herself from old habits, all the while paying homage to "the one thanks to whom she was there today".

Belloni's management was tumultuous and did not resolve any internal problems. Rather, she worsened them. For example, she harbored strong hostility against the director of the *Galaxie* and created tensions with other equipment managers. However, she had enough time to create a public company (SPL, *société publique locale*) to manage leisure facilities and infrastructure dependent on the city, seeking to better separate municipal activities from establishment management. Now renamed "Destination Amnéville", the structure also housed the tourist office and its staff.

In 2014, Éric Munier succeeded Belloni, winning against her in the municipal elections. He renewed his function for six years in 2020. The man was a lawyer at the Thionville bar. After measuring the internal tensions between structures and management (as well as the city's financial problems), he chose the slogan "to calm down, to clean up and restart". After being appointed to his office, he declared: "legal knowledge is essential in a lot of discussions [...]. I am not here to make myself rich, I am not looking for income or a springboard for a career [...]. The main goal is to keep investment projects despite a poor cash flow". He clearly focused on the real town and its inhabitants, without reducing the destiny of Amnéville to its tourist activities. "Management will have to be done reasonably. We are recovering from a situation where finances are the black spot. Before embarking on new projects, we will first have to clean up".

Regarding the transition between Kiffer and his successor, Munier made a harsh observation:

> the Kiffer era ended with his death. The transition with his deputy mayor did not work. We need to change our vision [...], to abandon disputes, find the solutions adapted to the expectations of the municipality and the requirements of each tourist structure. All my interlocutors are open people, determined to emerge from a crisis, engaged to Amnéville, eager to work towards a reset with clarity and confidence.

In fact, some establishments had financial problems (for example, the Snowhall), management difficulties or requested help for their renovation[11].

The authors of this work visited Amnéville for three days in August 2019. They met around 10 managers on site, as well as other actors in Metz and the *Center Parcs des Trois Forêts*. The scenarios were contrasting, rather favorable for the thermal establishments and the casino, complicated for the *Galaxie*, and difficult at the zoo, despite the large number of visitors. We could feel the nostalgia of a bygone great era and the hope for a city's revival. However, there was no entrepreneurial energy on the part of the town hall, nor, for that matter, any commercial or marketing initiatives. The focus was rather placed on general management, infrastructure (if necessary) and control, with the desire not to deviate from a conventional administrative framework.

At the end of 2020, with the Covid-19 crisis, the first layoffs arrived. The public company parted ways with a third of the tourist office's staff, at a time when it would undoubtedly have been more appropriate to relaunch marketing projects.

Contrary to Jean Kiffer's habits, the new mayor invoked management separation in relation to other threats weighing on the thermal center, as well as on a portion of the 2,500 employees working at the City of Leisure: "employers are the associations or structures that manage jobs [...]. I can only deplore the fact that there are layoffs".

---

11 Six years later (2020), once the Covid-19 crisis was over, the goals were the same, revealing that the situation was still challenging: "we must continue the work started: governance of the Galaxie, the work at the thermal center, the relabeling of the thermal and tourist site... We must continue to clean up the city's finances, which are always very delicate".

For its part, the zoo was taken over by an investment company. The latter replaced the historic director with someone who was supposed to turn things around and innovate. This person did not stay for long, though. Another replacement came in 2021, by a graduate promoting ordinary projects, imitating what large zoos do, without seeking to build a truly unprecedented initiative.

Following pressure to dismiss staff, the director of the thermal establishment resigned and slammed the door. The head of the *Galaxie* was invited to dismiss.

Nowadays, a large portion of the Kiffer system has been disassembled and resold to interested buyers. Among them, we can fear there will be predators (e.g. investment funds primarily concerned with financial profitability), or systems with conventional productive efficiency, which copy what already exists without seeking to be original.

The historic conglomerate may evolve towards another type of modality: a constellation of structures with their own decision power, which will require a minimum of overall coherence to ensure the destination's attractiveness, knowing that the internal itineraries and uses can diversify, depending on the visitors.

The aim of social paternalism will fade. It is not certain that the facilities are still dedicated to supplying jobs to the city's residents.

The purchasing operators each have expertise in their own type of activity, but with standardized management models. The role of "Destination Amnéville" could be reduced to infrastructure management, global communication and territorial development, provided that it generates clever and strong communication (which is not the case at present). Otherwise, the "spirit of the place" could be lost.

Service initiatives are also a source of worry. For example, the zoo is developing playing areas, improving its catering proposals, and bringing its shows in line with the times (bird scenography, as in many other recognized zoos). If Kiffer had lived, he would have made arrangements with Chinese entrepreneurs to have the pair of pandas before everyone else, including Beauval!

Today, Amnéville's pioneering period is over. At the town hall, it is the lawyer's language that predominates. All things considered, do we still have the energy to take action?

On the Municipality of Amnéville website, we can only see images of the town hall. As to the City of Leisure, there is only a summary description of the major attractions, without carrying much momentum:

Built over former slag heaps, a vestige of the steel industry, its tourist and thermal site, dedicated to well-being and relaxation, combines a variety of choices at the same place: water (with a thermal center and an Olympic swimming pool), snow (with the Snowhall, a unique indoor ski slope in France), nature (with a magnificent golf course and one of the most beautiful zoos in Europe), gaming and music (with a casino and a performance hall, the second largest in France). In addition to a wide range of activities, the thermal and tourist site offers outstanding catering and accommodation.

1.2. Center Parcs: a system driven by a large company

Center Parcs is a particular type of localized tourism system (LTS), built around an establishment depending on a powerful group. This company's head office is based in the Netherlands and belongs to the Pierre&Vacances Group, founded by Gérard Brémond in 1967. The latter imported the principle of vacation residences into France, that is, a set of residences (at least 100, according to French regulations) intended for tourist rental. The modality is as follows: first, the group undertakes a real estate development operation, builds the residences, then sells them to investors and finally facilitates them for rental management. Owners are individuals or companies who benefit from a tax reduction equal to VAT, provided that they agree to rent the residence for a period of nine years. At the end of the contract, a new period can be initiated.

1.2.1. *Brand history*

The principle of a resort[12] located in the countryside and close to a metropolitan area is older than the French model. This concept was gradually perfected. Center Parcs was founded in 1968 in the Netherlands. The initial idea was to offer the Dutch a formula combining comfort and relaxation close to large cities. Innovation was introduced by Piet Derksen, who had previously founded a chain of sporting goods stores, Sporthuis Centrum, whose first shop opened in Rotterdam in 1953. At the very beginning, he sold tents. Given their success, he then invented a bungalow formula which he deployed at a first campsite acquired in 1966, in Reuver (the Dutch Limburg). He designed it "with incredible luxury [...] for the time: sanitary building with saunas, main building with restaurant, dance room, TV room,

12 A resort is an enclosed tourist place managed by an operator who offers customers a global response to their needs, including accommodation, catering and leisure activities.

swimming pool with paddling pool, playground and supermarket…"[13]. In 1967, the entrepreneur moved onto permanent houses arranged in such a way as to limit the views. In 1975, five parks were installed. In 1980, he innovated once again with the creation of a new park, financed by the sale of the Sporthuis Centrum, within which the first tropical aquatic paradise was created amidst 600 bungalows in Eemhof in South Flevoland. In 1981, the group established itself in Belgium in Erperheide, then in De Vossemeren in 1987. The same year, he established a foothold in England at Sherwood Forest and in France at Bois Francs, in Normandy. Meanwhile, in 1985, the brand became Center Parcs.

The company was acquired[14] in 1989 by a Scottish brewing group Scottish & Newcastle, looking for an opportunity to diversify in a sector of consumption of its products. The smaller parks were closed, as the group established itself in Germany at Bispinger Heide in 1995, and the tropical paradise was renamed as Aqua Mundo. Then, in 2001, the continental residences joined the Pierre&Vacances group. In 2015, the six establishments set up in the United Kingdom fell under the group "Center Parcs UK", owned by the American company Brookfield Property Partners (Wikipedia, accessed on March 3, 2020).

The year 2022 was marked by an in-depth restructuring. The founder Gérard Brémont had to hand over the reins, while new shareholders entered the capital: "the British investment funds Fidera and Alcentra (approximately 20% and 25% each) become the new owners, associated with the French real estate asset manager Atream, which played a more industrial role". The triple principle of real estate development, sales to individuals and rental management, which made the group's fortune, had been shelved. In 2017, the allure of the proposal was undermined as the tax advantage granted to investors who rented their property for a period of nine years came to an end. Then, the pandemic also produced devastating effects. A total of 27 parks are part of Center Parcs Europe, a subsidiary of the Pierre&Vacances-Center Parcs group: seven are located in France (one unit opened in 2021), five in Belgium, six in Germany and nine in the Netherlands.

1.2.2. *The proposal*

Each resort comprises a set of residences located in a wooded park around a vast building, Aqua Mundo, which, in addition to an aquatic complex maintained at a temperature of 29°C and made of wave pools, houses a "wild river", slides and

13 "50 years of innovative holidays. Center Parcs together, for real", Irene Van der Linden, Robin Stam and Michel Haans, Krikke Special Books, in Opdracht van Center Parcs Europe NV, 2017.
14 Piet Derksen died in 1996.

various other elements (such as the "water tree", fun activities, restaurants and shops, particularly local products). It is a type of above-ground tourist location (MIT team 2005) in the sense that it offers a seaside and a tropical type of proposal close to the urban areas of north-western Europe, while sparing the constraints and fatigue of access.

The principle is to offer stays combining accommodation and access to partly integrated leisure activities less than two hours' drive from the metropolises of north-western Europe. Catering and other leisure activities are subject to an additional charge. The formula is attractive and benefits from a high fill rate. Center Parcs constitutes the flagship of the Pierre&Vacances group[15].

This assertion has been contested. Several projects are currently blocked or have been abandoned in France, such as in Roybon in Isère. The dispute mainly relates to environmental damage, because the company seeks wooded and relatively hilly areas for its locations. However, the creation of direct and indirect jobs constitutes a response to the desertification of the countryside. While the proposal attracts city dwellers who do not spontaneously move from the daily universe and urban work to discover a bucolic environment, the leisure proposals are not always attractive. Marc Augé (1997) has emphasized their artificial dimension, as compared to the "real" countryside. Although temporalities are different, the latter can be considered a social production, in that the resorts meet a real need.

The principle of clubs, implemented here, was invented in England by a stall holder, Jérémy Butlin, who tried to respond to the desperate wandering of tourists who stayed at a bed and breakfast, and were kicked out at 10 a.m. First, he provided facilities to accommodate physical activities and to promote sociability among individuals. It was a failure. He then supplemented his proposal with the presence of facilitators, responsible for encouraging idle city dwellers to become active. This innovation, in turn, was successful. In fact, for individuals to transition from the hierarchical and time-measured world of work (see Elias and Dunning (1995) and their reflection on routines), to long-term free time spent at a place that may very well be pleasant but unknown, is not obvious and requires mediation. The desire for the countryside, associated with calm and rest, comes up against the risk of idleness and boredom. A comprehensive response to tourist needs, the company offers accommodation and leisure activities to a population lacking the dispositions and skills (Guibert 2016) required to satisfactorily experience *deroutinization* (Elias and Dunning 1995) in the countryside.

The dispute also seems to be anchored in the illusion of a perfect world, without any effects on the environment. Of course, the resort involves a transformation and

15 Clement Guillou, *Le Monde*, March 10, 2022.

an occupation of the space, but therein lies the very principle of a production establishment. In addition, it is a coherent response to the need to escape routine, and this, within a short distance from Western Europe major cities. While proximity is considered as a virtue (in response to the environmental damage caused by transport, particularly air transport), this is not perceived in the case of Center Parcs.

1.2.3. *The Center Parcs actor system*

Center Parcs is an actor system characterized by a main company that builds subcontracting-type relationships according to two main methods. On the one hand, the group establishes bonds which are managed at the central level and shared by all establishments. For its part, restoration is given in concession to the Areas company, a former subsidiary of the Elior group which became independent after being sold to a pension fund, PAI Partners. It specializes in *retail travel*, the implementation of catering solutions that target mobile populations. It is therefore present in airports, train stations or motorway areas. It sets up points of sale in these sites and establishments that are franchises of different brands such as McDonald's, Brioche Dorée, Exki or Paul. Areas claims 150 brands. The choice for each "flow location" is made depending on the characteristics of the demand.

On the other hand, relationships are established with local companies, chosen by the local management of each Center Parcs. For instance, markets for local products are established through partnerships established on site. In Hattigny, in Moselle, it was the farmers who asked the management to be integrated.

Apart from this, there is a third method, less controlled by the company manager. Derived from the tourist inflow of each establishment, it is based on the secondary mobility of individuals who circulate from the resorts and undertake excursions in an area estimated at a 45-minute travel radio. For the *Bois aux Daims* branch, located in Vienne, the share of visitors exploring the surrounding area is estimated at 27%. About half had planned these outings before coming, while the other half designed the project once there. Non-residents tend to go out more. This is also the case for the wealthiest visitors and those choosing the longest stays. At *Bois aux Daims*, the department struggles to retain Center Parcs customers, apart from those in Futuroscope. There, Center Parcs is the most visited place (32% of respondents), ahead of Saumur (Maine-et-Loire, 24%), Chinon (Indre-et-Loire, 12%), the Fontevraud Abbey (Maine-et-Loire, also 12%) and Puy du Fou (Vendée, 11%), which are positioned far ahead of Poitiers (6%) or the *Vallée des Singes* in Romagné (5%). However, beyond the visits, excursions outside the park generate benefits in catering (for 46% of households out) and businesses.

Figure 1.3. *Communication from the Center Parcs company at "Bois aux Daims", which affected both the activities offered within the resort and the establishments and places located in the surrounding area (source: website, screen capture from July 22, 2022)*

As each tourist establishment is negotiated with legitimate actors, territorial imposition strategies are given free rein. In Vienne, at the Center Parcs known as "Bois aux Daims", even the local products distributed on site are sponsored by "Futuroscope".

The establishment of a group unit represents a significant volume of overnight stays (1 million for "Bois aux Daims", based on an average length of stay of 4.5 nights for 125,000 customers). This naturally boosts departmental tourism. For example, the economic benefits for Vienne were estimated at 4.5 million euros in 2017. Despite the efforts made to encourage individuals to circulate primarily in Vienne, the phenomenon also affects bordering territories, because boundaries are porous.

On September 16, 2022, the group, in difficulty, was recapitalized. It came under the control of three funds (25.4% for Alcentra, 24.2% for Fidera, 8.8% for Pastel Holding), to which creditors were added for 11.9%. However, Center Parcs is considered the firm's jewel.

1.3. Cognac: spirits giant, tourist dwarf

Cognac is an eau-de-vie and commonly referred to as a spirit. Its production is based on wine and is elaborated by distillation. Blending and aging are ensured at a vineyard of the same name, in the south-west of France, north of Bordeaux. Wine culture on the banks of the Charente dates to the 11th and 12th centuries when monks developed land plots, called “brandes”. Wine production was exported by barges, flat-bottomed boats which also transported salt to Rochefort, where the barrels were unloaded before reaching ports in Northern Europe. Altered by the sea voyage, the wine was then distilled by the Dutch in order to be made drinkable under the name of *brandy*, derived from *bandwijn*, which means undertow. However, its reputation was ensured by the British merchants who created several specialized houses in the small town of Charentes (Girard 2016): notably John Martell in 1715, James Hennessy in 1765 and the Scottish baron James O’Tard, whose grandson co-founded the Baron Otard house in 1795 (Coussié 2011).

1.3.1. *A product with a strong international reputation*

In 2012, 168.5 million bottles of cognac were sold, generating a turnover of 2.381 billion euros. That same year, cognac was the *leader* of French exports in the wine and spirits sector. Moreover, cognac is almost entirely sold abroad, with 97% of exports. This high rate can be explained by the history of cognac. The founders of the trading houses not only turned towards the international market in the 19th century but have also done so more recently. Since the beginning of the 21st century, these houses have sought to ensure growth on several international markets and to regulate production, following the crisis experienced at the end of the 1990s when the “uncertainties of world trade and political contentions hindered its [the cognac’s] development” (Bernard 1996). In 2012, the cognac industry recovered from this crisis. A 3.1% increase in sales (as compared to 2011) brought about a 16% additional revenue. However, a wine region does not automatically become touristy.

On the contrary, in comparison with this product’s global notoriety and the dizzying economic figures of the spirits industry, the tourist activity in the town and at the Cognac vineyard is far more modest. In 2013, the average length of a stay in the town of Cognac was only a day and a half, with the tourist office only receiving 40,000 visitors. This figure is far lower than the attendance at the Bordeaux tourist office, which recorded just over 842,000 visitors the same year. One of the explanations stems from the fact that the town of Cognac has a limited hotel supply intra-muros (351 rooms for 10 hotels on January 1, 2020) and an atomized offer of guest rooms and lodgings within the vineyard. However, hotels are not the driving force of tourism, but rather the general visits, which in turn induces investment in accommodation.

In fact, during an interview, an institutional tourism actor stated that "people don't know that there is a vineyard in Cognac. I don't know how they think cognac is made, but not with wine or vines in any case, they don't know that Cognac is a town, they see the product without knowing that it's a vineyard". On the one hand, multiple operators are involved in enhancing the vineyard's value by proposing enotourism, or spirits tourism. On the other hand, the vineyard actors are engaged in a process promoting tangible and intangible heritage. However, within this territory, the economic power and reputation of the large cognac houses occupy a considerable place, sometimes shadowing the tourism development actions undertaken by institutional actors.

1.3.2. *An existing tourist activity*

Figure 1.4. *Circuit departure point for visiting a trading house by boat in Cognac. Picture by Jérôme Piriou, 2013*

Strictly speaking, trading houses receive visitors without intervention from institutional tourism actors. According to the Poitou-Charentes Regional Tourism Committee, trading houses welcomed 55,966 visitors in 2012.

The intervention of multiple actors in enotourism even led the president of the tourist office, also vice-president of the French National Inter-professions' Bureau for the Cognac appellation (BNIC, *Bureau national de l'interprofession du cognac*), to declare in April 2014, during a conference for the representatives of the Charente tourism sector: "there is a strong need to connect the economic players in the region".

1.3.2.1. *Over half a century of public reception*

The main visit to Cognac is to the large trading houses. Six circuits exist in Cognac and one in Jarnac. These circuits are all designed following the same schedule. The first part is devoted to the presentation of wines and soils. The second part focuses on barrel-making and their export by barge. A final phase is dedicated to cognac distillation, a task during which the cellar master plays a key role. For some actors, the storytelling of this phase heavily relies on folkloric elements: as an institutional actor from Cognac suggestively stated: "there is a permanent flirt between the brand's history and the legend". The Baron Otard company, located in the royal castle of Cognac, claims to be responsible for the first opening to the public in 1963: "the intention was to promote the local know-how and to show pride in what the region was capable of doing at the time". Originally, the development of visiting tours was open to customers, "then little by little the [cognac] houses opened up, because there was a demand, and they offered visiting tours with increasingly wider ranges". However, the activity was not profitable, as its primary goal was to sell bottles: "our guides are not expected to act as museum tourist guides, but rather as ambassadors, wishing to share an experience around the brand". All the houses have renewed their circuits and overflow with imagination to differentiate from one another: "we always have the desire to stand out thanks to a little something, at Rémy Martin we take the little train, at Hennessy we take the boat (Figure 1.4), Baron Otard is set at François I's castle", an institutional actor from Cognac amusingly declared. However, behind a desire to showcase the production of cognac lies a brand strategy driven by a collective will. These large houses yearly reinvest 20–40% of their turnover in *marketing* and promotional support. The brand must be sold, particularly to target luxury markets. The AOC label (*appellation d'origine contrôlée*) thus fades into the background, as the product is difficult to perceive for its terroir by non-connoisseur foreigners. Small houses like Frapin – which also open their doors to the public – have more interest in developing knowledge of the product rather than the brand, which is less well known and will find it arduous to compete against the big players.

1.3.2.2. *A recent development*

However, we have observed an evolution over the past 10 years, alongside the desire to become part of a wine tourism project. In 2013, Rémy Martin received the national wine tourism prize awarded by Atout France, a sign of recognition, even though their target is not the tourist, but the consumer. Moreover, the term "enotourism" is controversial, because "for some actors, in the expression 'enotourism', the most important thing is tourism, whereas if one says that to the inter-professions' bureau, the most relevant part is 'eno', the product". This is not necessarily incompatible, as those responsible for visiting circuits at Martell suggest: "the idea is for visitors to come to Cognac, we target an international audience". At Rémy Martin, for example, the visit is considered as an activity in its own right:

> The management's goals are shared… The first goal is to be able to convey the territorial and family roots to the clienteles, for them to value the product's authenticity, but the second goal is ultimately to continue to develop circuits and experiences like a small business.

Beyond that, visits to these houses go well beyond the tour circuit, depending on the services purchased. Some can cost thousands of euros:

> The trading houses all have prestigious châteaux or outstanding guest houses in the vineyard, where they receive their clients. It is not ordinary tourism, they receive their customers, broadcasters, advertisers, importers, hoteliers: the focus is on customer relations (As reported by the Charente Departmental Tourism Committee).

For example, at Rémy Martin, we may purchase a one-day stay at the castle, served by a chef.

1.3.2.3. *Also present in cultural patronage*

We also see that cognac houses invest massively in the cultural sector and private sponsorship. For example, they financed the Cognac Crime Fiction Fair in order to lobby in favor of the drink by advertising scenes of cognac consumption in detective films. This initiative was abandoned due to lack of success. Therefore, they undertook other promotional actions, such as organizing the Cognac Blues Passions festival, where the houses invested a significant budget to invite loyal customers and offer privileged access, each house with its own VIP lounge. The town took part in its own way by providing a subsidy, but above all, by lending the Town Hall's public garden, which was privatized for the occasion. Other major cultural actions were implemented, such as hosting Picasso or Sempé exhibitions at Hennessy, classical music concerts at Meukow and even creating a corporate foundation at Martel in 2016, to welcome various artists who change regularly. At the end of the day, all these trading houses are differentiated by their blending of the final product, which contributes to the brand's specificity. In short, while the houses consider themselves complementary on the circuits, they are above all competitors on the products: As we were told at Rémy Martin:

> All the visits are interesting, there is certain complementarity to them. If we have visitors, we can send them to Hennessy, to Martell, without mentioning their names, but it also depends on whether we have visits to offer at the moment (or not). Our main interest is for the customer not to leave the territory.

The tourist office is considered an impartial actor when it comes to advising visitors.

1.3.2.4. *Cognac branch houses*

Trading houses may sometimes invite tourists to visit the winegrowers they work with. However, among the winegrowers who produce for their own benefit, there are only very few who offer tours. Moreover, they do not intentionally seek to do so:

> For the time being, our turnover is growing. What could really motivate us, is a crisis. The wine is selling very well at the moment, the price has been increased by the trading houses and everything is organized by the trading houses, but some producers know how to preserve themselves.

Opening a wine property involves not only welcoming the public, but also wanting to share the terroir, the product and the company's history. Unlike other trading houses, in the Cognac vineyard, producers are not single producers, but rather cultivate Pineau and Charentais wine in addition to cognac.

Other companies occasionally open their doors as part of group visits organized by the tourist office. This is the case of the glassmaking company Verralia (Figure 1.5) or the agricultural equipment company Grégoire, both located at Château-Bernard, in the Cognac agglomeration. For these businesses, openings enable them to raise awareness about the wine professions and the different sectors, and to counter the fact that the Cognac vineyard suffers from a lack of staff for these companies which require specific qualifications. The interest is weakly related to the territory's tourist economy, but rather to the agricultural and industrial economy. Other wine diversification companies are also seeking to make themselves known, such as producers of premium vodka or vinegar. For example, the Baume de Bouteville company, created in 1990, targets a different market, the food market, while taking advantage of local know-how in the production of raw materials, cauldrons and barrels. It opens its doors to the public, but under limited reception conditions, depending on staff availability and the site's security:

> We are enhancing our structure to welcome more groups and visitors. There is a sign on the factory's door to convey that the factory is open,

> but we do not mention any fixed trading hours, because we first focus on production (retail, boutiques, the distribution network, and catering).

The company also welcomes a few visitors sent by the trading houses and winegrowers, but their number is marginal compared to the groups sent by the tourist office.

Figure 1.5. *Glassworks for manufacturing cognac bottles, temporarily open to visitors. Picture by Jérôme Piriou, 2020*

The action taken by winegrowers for promoting tourism, particularly with "Les Étapes du cognac" circuit, is insignificant for the trading houses, when compared to the purchase of wine and the communication deployed on an international scale. While small trading houses understand the link between the territory and the cognac appellation, for the "big players", the BNIC's (*Bureau national de l'interprofession du cognac*) actions should focus on attracting talent, improving the image and providing an adequate infrastructure:

> The work of the BNIC, especially of the communication department, mainly addresses training… creating a school, an institute, a sort of academy, we agree on that. But, on the other hand, the visibility aspect in the town and developing enotourism is not our job, it is the job of the local actors.

However, there has been evolution in the BNIC's discourse, since as part of the development of a social responsibility strategy, certain territory-related aspects have been included. In 2019 and 2020, the inter-professions' bureau invited local actors (tourism, heritage, spirits, industry) to attend participatory workshops. Nevertheless, there is still a lack of articulation between the houses and the elected officials: "for example, the BNIC was not present for the regional season's greetings... there seems to be certain contempt from BNIC elected officials towards local officials".

Briefly stated, the houses do not need other actors to rely on but can work properly on their own: collective organization only functions in times of crisis. On the contrary, for the winegrowers who open their establishments for "Les Étapes du cognac" circuit, the purpose of tourism is not just to sell. There is indeed an emotional dimension coupled with the commercial one: "it [the circuit] is very popular among enthusiasts, as something to indulge in, but behind it there must be real economic impact". Those seeking to develop communication emphasize the importance of their vineyards, the grape varieties, their quality, the art of distillation; in other words, their sales pitch is rooted in the terroir and history. However, that approach is highly individualistic, as some producers do not know one another. The role of "Les Étapes du cognac" is to "show that there is a card that can be played with another member, to let people know what happens at each production stage". In Charente, "the walls are high". This expression reveals the desire of cognac producers to hide undeclared production and the traces of mushrooms on the cellar facades, resulting from the escape of alcohol vapor, "the angels' share". It also proves the compartmentalization between viticulture actors. Ultimately, even with an exceptional product such as cognac, little do the large trading houses with a strong presence at various shops in international airports (New York, Beijing, Shanghai, etc.) contribute for winegrowers to promote their territory, because the strategy is not the same.

1.3.3. *First obstacle: the weight of the large trading houses*

1.3.3.1. *Internationally oriented, trading houses alone at the helm*

As it happens in other vineyards, in Cognac, the distribution of activities is clearly established, following strict specifications to comply with the AOC label (*appellation d'origine contrôlée*) obtained in 1936. The winegrowers produce the wine, the trading houses purchase it, they control the aging of the *eaux-de-vie* and ensure their quality (matching requirements specific to each house). The BNIC (*Bureau national de l'interprofession du cognac*) is the institution responsible for promoting the product on an international scale. It plays a major role in fixing the prices of the grapes that will be commercially exploited. No less than 4,255 farms work under exclusive contracts (or not), for 277 traders. Each trading house has its own alcohol distillation and aging system (Bernard 1996). Furthermore, the most powerful among them manage a large part of the production chain, from contracts with the winegrowers to the management of sales points and the marketing of bottles around the world. Even though there are many houses, the cognac market is highly concentrated, since 80% of sales of cognac bottles are made by four large trading houses (Hennessy, Rémy Martin, Martell and Courvoisier). The crisis experienced at the end of the 1990s led many houses to join large international luxury groups or in the wine, spirits and alcohol sector. Hennessy, already merged with Moët, joined LVMH. Martell joined Pernod Ricard. After joining forces with Martini & Rossi, Baron Otard was included in the Bacardí group. More recently, Courvoisier (located in Jarnac) was integrated into Beam Global Spirits & Wine. Also, the main decision-making centers have relocated from the Cognac vineyard to Paris, New York and even Bermuda. This trend signaled the end of the massive local investments in vogue since the 19th century. According to an institutional actor, these houses "are nowadays disconnected from the life of the town (of Cognac). The activity of these houses is based on repositioning them in production and marketing". On the contrary, an entrepreneurial actor states the opposite: "here are major private players, large companies with their own interests, for which the public policy perspective is often quite disconnected from that of the companies".

To explain the trading houses' strategy, a local actor testified: "large houses need neither a cluster, nor heritage recognition, nor tourism". Moreover, the actors of tourism are well aware of the situation: "trading houses have a market presence backed up by their capital, their target is not the tourist, but the consumer, that is why we sometimes have problems with communication". Moreover, the BNIC is well aware of the power of branding:

> Cognac is a vineyard where the appellation is almost erased by the commercial success of the big trading houses. Consumption at an international level mainly points to Rémy Martin, Hennessy, Martell, people know it is cognac, but hardly do they know that cognac is also an appellation or a town.

For example, the person in charge of the visit circuit at Martell confirms the brand's interest:

> We are now negotiating with Russian, Chinese and American tour operators. The idea is to make people understand where Cognac is. When I was asked to take care of Martell, I had the address book of all the operators who work with Mumm, and they found it interesting to develop an action in the town of Cognac.

The company managed the contacts directly, without the intervention of traditional tourism actors. According to the BNIC, the houses welcoming hundreds of Chinese visitors without making the territory profit are using the wrong strategy:

> They [the territorial communities] say we are there for the product, for the territory. But if they are not present on the territory, in the long term, they will be discredited for international receptions. They cannot be legitimate on our territory if they are not there.

Truth be told, with a budget of 2.4 billion euros (surpassing the Champagne inter-profession in 2013), the inter-professions' bureau is mainly financed by the trading houses, in particular, by Hennessy, the leading firm.

1.3.3.2. *A few of the initiatives by small houses and winegrowers*

During one of the city enhancement sessions organized in June 2017, 14 professionals with a diverse background – hospitality sector, tourist office, consulting, air force, heritage association, aerodrome, catering, university, industry, Chamber of Trades, EPCI (*Établissement public de coopération intercommunale*) – discussed some of the actions to be carried out:

– the launch of an association to inventory the intangible heritage of the Cognac vineyard, "cognac's know-how", with a view to applying for the inclusion on UNESCO's World Heritage List;

– avenues for improving the accessibility conditions to the territory by road, rail and air;

– partnerships with public institutions (notably with the EPCI establishments) to facilitate recruitment and reception in the territory;

– creation of a welcome guide for companies and employees.

During the cluster's first three years, these sessions made it possible to engage different actors:

> Working helped us to clear the landscape. This sketch made it possible to outline the territory, make it sustainable, determine what we were going to build on it, outline a scenario, roughly get to know the forces involved.

However, participation was random and tended to decrease over time:

> At the beginning there was a lot of enthusiasm, everyone contributed their view and the exchange quickly turned into *brainstorming*, everyone practically said the same thing every time. Then the interest faded a little. This was problematic and no apparent solution arose.

Furthermore, the arrival of new members required adapting the format. Coordinating a community of around a hundred members, with only one cluster director and two project managers, is a particularly perilous exercise. Also, a certain form of autonomy was launched around project workshops in order to better target participants (examples of themes: creation of a platform bringing together job offers in the region in relation to the spirits ecosystem; design of a guide to economic attractiveness; writing of an article for the specialized press). However, again, the question of numbers arose; we said that there should not be more than three to four persons, but each time there were a little more people who positioned themselves.

Furthermore, representatives of major cognac trading houses took part in the discussions and were often the driving forces. However, the process of gathering various actors in the pursuit of a collective project is sometimes complex. "The goal is not to come with preconceived ideas and impose them in a discretionary manner". On the contrary, the largest cognac trading house was not a member of the cluster. The "Spirits Valley" cluster had to develop a persuading force and find its place:

> We have positioned ourselves, this is not only strategic, but also political. Compared to an environment with existing clusters, compared to the BNIC, to the Chambers of Commerce and Industry, finding our place in this ecosystem, our place as a catalyst, our place as a unifier, is a daily job. We are starting to have legitimacy, but it involves long-term work.

Finally, at the beginning of 2020, among the members of the "Spirits Valley" cluster, a core of companies and organizations renewed their membership. However, for others, it is more complicated to devote time to the collective, especially as an inter-professions' bureau already exists for the vineyard, and product diversification by the actors (vodka and various spirits) does not make up part of its current policy.

In the vineyard, the association "Les Étapes du cognac" helps to promote tourism in viticulture-dedicated sites. At the end of the 1990s, a general councilor of the Ségonzac canton brought the idea of a group of young farmers to create a tourist route, as the one already existing in Alsace. While in other vineyards, it is the inter-professions' bureau that coordinates this type of activity; this is not the case in the Cognac vineyard. The local communities, in particular, the General Councils of Charente and Charente-Maritime, have taken up the reins. In 2000, a circuit was created, called "Les Étapes du cognac", because the name "La Route du cognac" had already been registered by a communications company from La Rochelle. The association is organized around three colleges: the financing agents (departments, the region, trade unions, the inter-professions' bureau), the institutions (tourist offices, departmental committees and regional tourism committee) and the members, who are mainly winegrowers (173 members in 2013). The association has four main missions:

1) support members on a voluntary basis for improving their structure to create a wine tourism offer within the farms (creation of shops, discovery trails, etc.). The particular trait of the Cognac vineyard is that 95% of winegrowers work with the trading houses, direct sales constituting only a small part of sales;

2) propose entertainment, by organizing summer activities;

3) compile a directory with the support from the Cognac Experience platform of the Charente Departmental Tourism Committee;

4) network and facilitate exchanges among winegrowers regarding the supply's complementary nature.

The association is financed up to 17% by annual contributions from members based on a forecast budget of approximately €150,000. The complement is provided by various actors. The most important financing agent is the Cognac inter-professional association (18%), then Pays Ouest Charente, who pays its subsidy via the General Council of Charente (13%), then the General Council of Charente (12%), the General Council of Charente-Maritime (8%) and finally the county of Haute-Saintonge and the region of Pons. In 2010, the association received the "Vignobles & Discoveries" label delivered by Atout France to qualify the wine tourism offer, based on an evaluation grid for each actor category (winegrowers, hoteliers, catering, etc.). A first sign of alert came in 2013, when local authorities reduced the amount of the subsidies to the association. They expressed their desire to disengage:

> It is also a way of inviting the inter-professions' bureau to take over, the local authorities think that it is not up to them to pay for a cognac which is doing very well… they might pay for it if there was a crisis, but these are also political reasons and we are in the middle of all this.

Finally, in October 2019, the budget deficit meant the end of the association.

1.3.4. *Second obstacle: an organization of the sector essentially oriented towards the production of wines and alcohols*

In 2010, a university study carried out for the DREAL (*Direction régionale de l'environnement, de l'aménagement et du logement*) of Poitou-Charentes and the

DDT (*Direction départamentale des territoires*) of Charente revealed the existence of a group of activities resembling a cluster:

> The geographical concentration of complementary activities – the establishments in the territory specializing in each of the production stages of the Cognac *eau-de-vie* – a priori reveals the existence of a cluster (Bouba-Olga et al. 2012, p. 182).

However, the creation of trading houses during the 18th century made it possible to develop an industry around the production of Cognac within the vineyard itself. At the same time, paper production was established there thanks to mills and industrial machines. There was also the manufacture of barrels and glass, notably through the creation of a quasi-automatic glass molding machine invented by a Cognac glassmaker at the end of the 19th century (Bouba-Olga et al. 2012). Then, we must be able to market the Cognac, transport it, store it. By 2012, there were an estimate of between 13,500 and 16,000 jobs directly or indirectly related to the territory's production: "75% of the agglomeration's economic activity is linked to its production" (Bouba-Olga et al. 2012, p. 182). Some companies were in charge of manufacturing wooden packaging, corrugated cardboard, hollow glass and printing materials. The production of Cognac requires numerous activities around cooperage, glassmaking, corking and even packaging.

In addition to the presence of actors involved in the production of Cognac, the presence of a cluster can both be explained by cognitive factors and transaction costs (Bouba-Olga et al. 2012, p. 186):

> – A local concentration of related activities for companies to reduce transaction and transport costs caused by distance: Saint-Gobain (a glassmaking company) strategically settled in Cognac, close to the trading houses and wine properties.
>
> – The production of *eaux-de-vie* requires a meticulous know-how, with specifically related activities within the Cognac vineyard. The Vicard cooperage, established in Cognac since 1925, produces 40,000 barrels per year, adapting production to the houses' expectations, particularly concerning the composition of the wood favoring aging, in particular for the oldest *eaux-de-vie*, kept in small barrels called "dames-jeannes" carefully stored in the cellars' "paradise".
>
> – A training program, specific to the region, intimately related to viticulture and enology, justifying the presence of a university campus in Ségonzac and even a cooper school in Cognac.

– Social relations between the actors have fostered international development. Territorial anchoring is key to the AOC label (*appellation d'origine contrôlée*), because it delimits the vineyard and specifies the conditions for producing Cognac complying with strict specifications, which also justifies the choice of location for the industries.

The constant sales growth in Cognac bottles prompted the trading houses to encourage winegrowers to plant more vines. On an area of approximately 75,000 hectares, 250 hectares were authorized in 2016, 800 hectares in 2017, 1,557 hectares in 2018 and 3,474 hectares in 2019. This authorization is subject to deliberation by the Charentes-Cognac wine region council under the chairmanship of the regional prefect, and validation by the French National Institute of Origin and Quality (INAO, *Institut national de l'origine et de la qualité*) and FranceAgriMer. Afterwards, an inter-ministerial decree establishes the limits so that winegrowers can submit individual requests. The Cognac industry is likely to continue growing in spirits sales, even though some winegrowers fear a reversal of circumstances, remembering the oil shock of 1971 and the crisis of the late 1990s. Other actors in the Cognac industry are joining forces in order to consolidate sales and maintain an international positioning for the star product from the Charente region.

1.3.5. *The creation of clusters in the Cognac vineyard*

In June 1998, in a gloomy economic context for the cognac industry, a handful of independent entrepreneurs in the packaging sector grouped their forces under the leadership of the Angoulême Chamber of Commerce and Industry (CCI). It was the director of economic and international action of this CCI (also the general delegate of the French National Institute of Packaging Design), who launched the "Atlanpack" cluster with a desire to create a "Spirits Valley", borrowing the concept from the American Silicon Valley. Originally limited to the Poitou-Charentes region, the cluster opened to neighboring regions in 2002 in order to integrate a total of around a hundred companies, 60 of which are located in the Cognac vineyard. With an annual budget of €450,000, 30% of which was self-financed, the rest of the subsidies came from Europe, the French State, the Department of Charente, the Cognac agglomeration and the Chamber of Commerce and Industry. The packaging sector (which not only works for the cognac industry, but also for the agri-food or pharmaceutical industries) benefits from this cluster of companies. Being awarded a certification (a "label") by the Ministry of Rural Space and Territorial Planning in 2010, it promotes the region's know-how and network coordination. Training on careers in packaging (both at high school and degree level) was proposed from 2003. It is dictated at a high school in Angoulême, and there is also a master's level training at the University of Poitiers. Certified as a cluster in the Nouvelle-Aquitaine region in 2016, Atlanpack was limited to 62 members in 2020. Every year, a forum brings together professionals in the sector.

In 2010, a study was carried out for the DDT of Charente and the DREAL of Poitou-Charentes. The resulting report was entitled "Cognac, analysis of a cluster of activities" (Bouba-Olga et al. 2010). The above-mentioned study was relayed by the mayor of Ségonzac (appointed to the regional council) and was carefully read by the Charente Chamber of Commerce and Industry. In 2013, a new study was commissioned by a research firm. In 2015, a new association was created, the "Spirits Valley", whose goal was:

> To ensure the promotion and influence of the know-how of the cognac sector by local actors, to develop and produce *super premium* spirits in collaboration with all the actors involved in the value chain, with a view to promoting and stimulating the growth of the local economy.

This association was not responsible for exclusively bringing together the actors in the cognac sector. This project engaged all the actors in the "premiumization" of the cognac value chain. This included the corkers, the labelers, the coopers; in short, all the different services required for other spirits such as Grand Marnier, liqueurs or vodkas. This cluster differed from the BNIC (whose role was to regulate the AOC sector) by a desire for networking with all the operators in the spirits chain. Governance was mainly ensured by public actors. In 2017, the board of directors was exclusively formed by local or consular appointed officials (president and vice-president of the Departmental Council of the Charente, Charente departmental councilor and president of the tourist development agency of Charente and Charente-Maritime, mayor of Cognac and vice-president of the Charente Chamber of Commerce and Industry and president of the Cognac delegation). Its president was an industrialist, elected to the Charente Chamber of Commerce and Industry. In 2020, this board of directors diversified, incorporating actors from the industrial, academic and even transport fields. During the launch phase, at the beginning of 2016, elected officials injected an operating investment of €250,000 into the association (financed by the Charente Chamber of Commerce and Industry, the Department of Charente, the European Union, the French State and the Poitou-Charentes region). In January 2018, the association opened a membership phase with an amount indexed according to the turnover of the adhering organization. In June 2019, the association had 100 member companies and institutions. At the very beginning, between 2016 and 2018, four to six working sessions were organized yearly around four axes: economy, training, attractiveness and logistics.

In 2016, the incubator of the Cognac branch of the Charente Chamber of Commerce and Industry welcomed the creation of a local branch of the

"Inno'vin" cluster, created six years earlier at the *Institut des sciences de la vigne et du vin* (ISV), located near Bordeaux. Led by an employee of the Charente Development Agency – the economic development body of the Charente Departmental Council – the goal was to provide those involved in the Cognac vineyard with an exclusively wine-producing cluster. In 2016, the "Inno'vin" cluster had an annual budget of €380,000, half of the funding coming from members and the other half from the local authorities. In the Cognac vineyard, coopers who already worked extensively with Bordeaux winegrowers were members. Then, they were joined by cognac houses, various industrialists (vat manufacturers, etc.) and scientific and university laboratories.

In 2017, the Grand Cognac agglomeration, then chaired by the mayor of Cognac, launched the project of a Cognac Trades Center, by purchasing a site from the Hennessy company. However, in 2018, the project became more ambitious and gained momentum by engaging cooperage, glassmaking and trading actors. The desire was to create a center which would bring together business incubators, clusters and economic development services from the EPCI. At the same time, the "University of the *Eaux-de-vie*", a university center located in Ségonzac, benefited from funding for works to increase its activity and receive more students. The aim was also to respond to a strong demand for labor and the need for specific qualifications to meet the expectations of the cognac industry.

The landscape became not only marked by the extension of vines, but also by buildings and cognac's characteristic industry. Indeed, there were cooperages, glassworks, but above all cellars (recognizable by the stone's blackness, caused by a microscopic fungus which settles on the vaults of the buildings where barrels age). The cellar master's role is to ensure the production of cognac, a blend of *eaux-de-vie* resulting from double distillation. Large and smaller trading houses control the production, particularly viticulture, and market several types of cognacs, which vary depending on their aging degree (Piriou 2018b).

1.3.6. *The instability of institutional structures*

Over the last five years, tourism institutions have undergone profound transformations, associated with two main reforms. First, the law of January 16, 2015 relating to the delimitation of 13 large new regions, replacing the 22 existing metropolitan regions, without modifying their integrating departments. Then, the law of August 7, 2015 relating to the new territorial organization of the Republic, known as the "NOTRe" law, which modified the institutions' competences. Regions became leaders in economic development, while the departments maintained an orientation as guarantors of human solidarity and territorial cohesion. As for the EPCI (*Établissements publics de coopération intercommunale*) establishments, the

NOTRe law amply modified their attributions – particularly the tourism competence – with a compulsory transfer from the municipalities, except for those classified as tourist resorts. In addition, the minimum thresholds went from 5,000 to 20,000 inhabitants, except for mountain areas and sparsely populated areas. This means that the Cognac vineyard experienced notable modifications as to its institutional structures, making it possible to further clarify the role of the different actors and to position the territory as a tourist destination.

1.3.6.1. *At the local level*

The Cognac tourist office, with associative status and created in 1924, has transformed its scope of action over the last 20 years, in order to offer a territory more in line with the vineyards in the Charente Department.

1.3.6.1.1. Grouping and scaling

In 1999, five tourist offices in the west of the Department of Charente (Châteauneuf-sur-Charente, Cognac, Jarnac, Rouillac and Ségonzac) signed an agreement to pool their actions with the aim of creating a tourist center extending across the Ouest Charente region, under the territorial coherence and orientation plan. In 2012, the regional policy of the Regional Council of Poitou-Charentes favored the creation of this Cognac region center, without legal status, and entrusting coordination to the Cognac tourist office. As of 2014, an annual agreement of goals was concluded with the community of municipalities of the Grand Cognac over a period of three years. However, it was the NOTRe law of 2015 that led to the merger between the five tourist offices in 2017, after creating the new Grand Cognac agglomeration (which integrates them and collects the transferred tourism competence). A local Tourism Development Plan was set in motion, defining the axes of the territory's strategy for the period 2017–2019. "Destination Cognac", a shared brand registered in 2017 for the communication and promotion of the tourist office, relies on a solid tourist supply, as well as four reception offices located within the noblest AOC wines. In 2017, the year of the merger, the tourist office, still under the associative status, had 362 members (80 additional ones compared to 2016). The board of directors had five colleges (12 members in the agglomeration college, nine members in the trading and viticulture houses college, four members in the inter-professional structures college, eight members in the college of professionals and representatives of cultural and leisure tourism; and finally, eight members in the college of professionals representing catering, commerce and accommodation). We may also note that 12 members came from the wine and spirits sector (nine members from trading and viticulture houses and three members from inter-professions' of Cognac, Pineau and Charentais vineyards). Together with the elected officials, these members constituted the second representative force on an equal footing. Furthermore, the four largest trading houses were represented (Hennessy, Martell, Rémy Martin and Courvoisier). However, the

Camus and Meukow trading houses, present in this college in 2016, were replaced and supplemented by Maison Boinaud, Cognac Lascaux, Guillon-Painturaud and Fussigny houses. The *marketing* power of Cognac brands is such that the tourist office also promotes urban heritage as a complement to company visits:

> Tourists think that cognac is made in a house, so there is a lot of communication work to be done, that is why we are focusing on the town's heritage, because for me the product is closely related to the town. We entice visitors to economic discovery through heritage guided tours, which does not prevent us from selling visits to bottles or cooper manufacturing areas.

The tourist office has a guiding service as well as an incoming travel agency (the only one in Cognac for over 30 years), which receives 300 groups per year and 8,000 people, selling trips, day stays or à la carte proposals along sites associated with the cognac industry (barrel-making, glassmaking, etc.).

1.3.6.1.2. Achievements

In January 2012, the town of Cognac obtained the "Town and Region of Art and History" certification, conferred by the Ministry of Culture and Communication. This distinction was awarded for having emphasized the territory's values and the town's identity around cognac: "we wanted to give a prominent place to cognac, but due to the fact that there are very few vineyards in the town of Cognac itself, it's not that we don't talk about it, but we do not bring it forward in the official file". The heritagization of the Cognac vineyard is specific, as it encompasses both the cultural and commercial dimensions.

> This is what makes the town of Cognac a kind of conjunction between a land where wine is grown, where wine is totally suitable for distillation, where most of the distillations are carried out on the vineyard, and a town with a river, a commercial vector for the merchants and traders in the city.

Cognac has this central role in the systemic organization of cognac production, and the valorization of its heritage by the town is part of the strategy. Once the roles of distillation and of the cellar master have been explained (through a tour of the trading houses), the heritage service evokes the industrial architecture of the cellars, the civil architecture of the cellar masters' private mansions and that of the big banks, testifying to Cognac's flourishing economy in the 19th century. Contrary to this, the trading houses do not promote the town's heritage, but their own products and brands.

Furthermore, heritagization reveals a "paternalistic" system maintained by the trading houses that goes beyond simple wine production. Many big names in the trading houses successively took turns as mayors of the town (Frédéric Martell, Jean-Baptiste Antoine Otard, Gabriel Martell, Paul Firino Martell, etc.), and until the 20th century, half of the members of municipal councils were merchants or winegrowers. From then on, many decisions taken for the town were also compatible with the interests of their businesses:

> In the 19th century, Cognac was well-organized to set up a water supply system in all the town's districts. As good as it was for the inhabitants to have access to water in the town, it was especially interesting for the trading houses to have water in the cellars.

Also in the 19th century, the town equipped itself with sports facilities, as the trading houses made significant investments in different sports (rowing, rugby, fencing). Today, the heritage department of the town of Cognac continues its promotion activities, through guided tours. It also wishes to set up more partnerships with large houses (particularly at the financial level) to organize awareness-raising activities for residents and tourists.

1.3.6.2. *At the departmental level*

Charente had its own departmental tourism agency until 2017. In 2006, the Department of Charente developed a partnership with the Departments of Charente-Maritime and Gironde to offer a joint supply for tour operators via the "bordeaux-cognac.com" web platform, an operation that then came to a halt. Then, in 2008, a consumer communication called the "Cognac Experience" was launched at the request of the General Councils of Charente and Charente-Maritime, each having invested approximately €50,000, in order to boost communication of the Cognac vineyard. In its 2011–2015 Charente Tourism Development Plan, the Departmental Tourism Committee aimed to continue the actions initiated by the Departmental Tourism Committees of Charente-Maritime and Gironde, the BNIC, Cognac's Tourism Office, the Charente Chamber of Commerce and Industry, in particular, through the websites *Expérience Cognac* and Bordeaux Cognac. The strategy aimed to be part of regional policy promoting the "Cognac destination". The principle of "Cognac Experience" was to create a tourist supply fostering the visibility of the Charente vineyards destination. The goal was to promote tourism in the wine industry and to improve the supply via the association "Les Étapes du cognac". At the time, the BNIC's communications director was greatly in favor of this initiative. The departmental tourism agencies then took control of this project.

The persisting concern was bridging the gap between the product's image and the tourist supply. Tensions also arose between the world of tourism and inter-professions' bureau, in particular, after a communication issued by *Expérience Cognac* which associated the Charentaise (a type of slippers) with barges and wine.

The problem came down to the fact that popularization was incompatible with the image conveyed by the cognac houses for the sale of their products. At the Charente Departmental Tourism Committee, there is a saying that:

> Cognac is not receptive, but deceptive: in relation to the image conveyed, that the product suggests, super high-end. There is not enough high-end accommodation, there are Bed and Breakfast facilities, but Cognac's international clientele does not stay in that kind of lodging.

However, in 2015, the departmental elections tilted the Department of Charente towards the same political side as the neighboring department of Charente-Maritime. This context would then contribute to the merger of the two Charente Committees under a single entity.

1.3.6.3. *At the regional level*

"Cognac-Vallée de la Charente" was integrated into the strategic policy of the Poitou-Charentes Regional Tourism Development Plan (2007–2015) as one of its tourism brands. The mayor of Cognac was also the president of the Regional Tourism Committee of Poitou-Charentes until its dissolution in 2013, requested by President Ms. Royal. The staff and actions were integrated to the services of the regional council, to enhance attractiveness.

As part of the regional organization reform, Poitou-Charentes was integrated into a new and vast entity, Nouvelle-Aquitaine. In May 2017, the members of the Regional Tourism Committees of Limousin, Aquitaine and Poitou-Charentes teamed up to create a single entity, entirely dedicated to Nouvelle-Aquitaine. The association's budget amounted to 6 million euros (three times more than the former CRT Poitou-Charentes) and ensured the promotion of tourism, coordination of institutional actors, offered engineering and economic observation services and finally took part in the development of the Regional Tourism and Leisure Development Plan. This plan for Nouvelle-Aquitaine, adopted in June 2018, identified Cognac as one of the brands to position internationally, or to strengthen on a national market. Furthermore, Cognac was also recognized as a "rural destination identified and structured for tourist clienteles" (ibid., p. 28).

The creation of the Nouvelle-Aquitaine region also contributed to this rapprochement decided by the presidents of the departmental tourism agencies of Charente and Charente-Maritime. A new association "Charentes tourism" was born in

June 2017. This new association considered the Cognac vineyard as a strategic territory for international clients, since it had applied to obtain a "Cognac vineyards" destination contract. This system supported by Atout France allowed the territory to benefit from financial aid of €75,000 and an action program over a period of three years. This application, sponsored by Charentes Tourism, was a compilation of a collective file with around 20 partners, including EPCI vineyards, the BNIC and the major cognac houses. The collaboration worked but remained limited to tourism promotion.

1.3.7. *The luxury trap*

As long as cognac is positioned as a luxury item, tourism can neither respond to it (due to a lack of adequate equipment), nor develop within that niche.

As for institutional actors, they claim their mission is to promote the territory: "for us, our job is the territory, it is not to sell the product, even if we are lucky enough to have a product carrying the name of a town". What is also at stake is that before going to the vineyard, visitors have a "high-end" image of the territory, which they associate with cognac brands. The director of the tourist office witnessed the return of the "Rendez-vous en France" trade fair in March 2013, where Russian customers requested services including massages and visits to trading houses with meals: "we have been identified as a niche destination, we are very high-end. So, everything is done on a product that is very emblematic, but which is very difficult to sell". According to information from the tourist office, individual customers come to tour the town, visit a trading house, and possibly take a boat trip. However, the town of Cognac is a place of passage; tourists only stay for half a day.

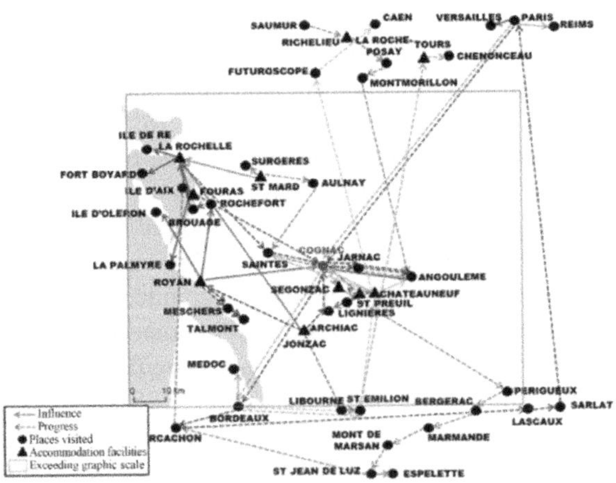

Figure 1.6. *Trips of the 10 tourists interviewed at "François I Square" in Cognac on May 10 and June 14, 2017. Jérôme Piriou, June 2017*

COMMENT ON FIGURE 1.6.– *Among the tourists interviewed, seven were French, one English, one American, one Chinese. Accommodation facilities are located in the Departments of Charente and Charente-Maritime, near Cognac, Royan, La Rochelle and near Surgères. There are three visiting areas: Charente (Cognac, Angoulême, Jarnac), the coast (Royan, the islands, La Rochelle) and the southwest of France (Bordeaux, Périgord, Basque Country). However, some tourists set the visit to Cognac within a more complex route, including the Loire Valley and the Paris region.*

The most important thing to note is the visitor's place of origin. According to the tourist office, flows come from the coast, especially when the weather is gloomy, as well as from Bordeaux and Dordogne. Others integrate Cognac as a stopover along a route between the north and the south of France (Figure 1.6). As for foreign customers, in 2013, the tourist office started welcoming Chinese visitors in family groups, who were looking for luxury services. Unlike Paris or Bordeaux – the preferred destinations for this clientele – the Cognac vineyard could only offer them to a limited extent.

1.3.8. *The legitimacy of piloting the governance of actions in the "tourism/spirits" couple*

According to Atout France, in 2016, the Cognac vineyard constituted only 4.4% of the French wine tourism offer, compared to 18% for Bordeaux and 17.2% for Champagne. This organization promoting French tourism internationally created an "enotourism cluster" in 2000, which had previously been called "Club Vitivini", bringing together partners, institutions and French wine professionals from 17 French vineyards within a voluntary and collective frame in order to develop enotourism. The multiplicity of actors in the cognac vineyard evidences the difficulty of defining the legitimacy of an actor. In 2013, interviews conducted at the BNIC and at the Cognac Tourist Office showed that both structures had a prominent role. Furthermore, the Rémy Martin firm, which won the enotourism prize in 2013, strongly believed in the unifying role of this cluster:

> The enotourism cluster is developed by dynamic people. At Atout France, the facilitator is someone very dynamic who knows their sector in depth and also knows how to pull the strings with the public as with the private actors. When all these questions have to be tackled, meetings are very dynamic and take place with real subjects.

However, on a vineyard scale, an enotourism cluster still struggles to structure itself. This is quite sad for a territory where the cognac and spirits sectors have been institutionally developing clusters for 20 years and whose state of mind has been present for centuries, as evidenced by an actor involved in the premium spirits cluster, as well as at the tourist office:

> This cluster operation has been taking place for 3 centuries now. It became formalized within an institution (a unifying organization) only 2 or 3 years ago. I took part in the genesis of this organization, I found it logical to structure certain subjects, to rely on synergies between the companies to work on subjects such as logistics, infrastructure, the region's allure, tourist appeal, but most importantly, employee attractiveness. This targets the customers, the companies who come for setting up their businesses, it all comes down to simplifying the structure and increasing the quality of life.

1.3.8.1. *The spirits economy dominating the tourist economy*

Everything has been said: in the vineyard, it is easier to create a cluster in the interest of the cognac and spirits sectors, to respond to industrial interests (recruitment, development, innovation) than to promote the tourist and heritage sectors. Something similar happens on the BNIC's side, where members openly admit that tourism is not the priority of their actions:

> There is a *gap*, at the collective and at the territorial level. However, very quickly, I saw, and even here internally, that our structure is not yet solid for enotourism, nobody has fully engaged in that direction.

This is also explained by the fact that although some of the trading houses would be in favor of tourism, others, particularly among the largest houses, remain far more reluctant. The interest of these houses is not Cognac: "it's not their playing area, [in fact] it's China and the United States".

Initiatives were launched for the tourism sector, starting with "Expérience Cognac", which in 2008 brought together the departmental tourism committees of Charente and Charente-Maritime, the BNIC, "Les Étapes du cognac" and the tourist office. The problem is that this initiative has not achieved consensus. One of the member actors claimed: "here we say that it was us who asked for Cognac Experience, and for their part, they say that it was a request issued by them". Another actor believed that:

> Expérience Cognac is a vast nebula, merely the toy of an institution. This cell would rather make sense on the international map because it is the Regional Tourism Committee who is in charge of the brand 'Destination Cognac-Vallée de la Charente.' They claim they work on communication at a national level, but that doesn't make sense with what we want.

For the tourist office:

> It is necessary to redefine what everyone does in the territory, people have to define, people have to know… For me it is clear, we have 'Les Étapes du cognac' which qualify and animate, we play to the music and the BNIC takes charge of professional training and communication on cognac, I would say we have the infernal trio.

Furthermore, at the beginning of the 2010s, the idea of a cognac cultural center was born, whereby the BNIC would be in charge of the training and the tourist office of the "window display": "a joint supply, without getting married in the strict sense, but sharing premises and a common communication strategy". For its part, the heritage department of the town of Cognac even encouraged this project:

> It is relevant and even neutral in relation to each of the trading houses. We even said that as Cognac had been labeled as a town of art and history, we should create an interpretation center for architecture and heritage, hosting a documentation center, an educational space, permanent and temporary exhibitions. At present, this does not exist in Cognac.

Moreover, the heritage department of the town of Cognac successfully introduced a panel into the tourist office presenting cognac in all its forms (architecture, cellars, houses), at a place where, until then, there had only been an exhibition of cognac bottles.

1.3.8.2. *Sharing the "cognac" brand for promotion*

In 2016, the relations between BNIC and the tourist office became strained over the communication around "cognac". In December 2015, the tourist office registered the "Cognac spirit" trademark with the French National Institute of Intellectual Property. This brand was used on various products in the store (T-shirts, caps, umbrellas, bags, etc.). For the tourist office, the expression "Cognac spirit" meant "the spirit of the town of Cognac". However, the BNIC did not perceive it from the same angle, but rather as "cognac, the spirit" (in the sense of the drink). Having given its consent during the summer of 2015, the inter-professions' bureau took its decision back in November 2016, considering that the term contravened the appellation's protection. According to the inter-professions' bureau, the trademark registered by the community of municipalities risked being purchased. Therefore, the BNIC bought it from the tourist office to ensure that a third party would never have access to it. The communications director nuanced the tension between the inter-professions' bureau and the community of municipalities via the tourism office: "we took the time to ponder about this case, we reacted very quickly. We all want tourism to develop, and we are working together to develop it".

At the end of the 2010s, the BNIC's position changed on enotourism, and more particularly, as to its place in the territory, no longer falling under the appellation. This strategy also made it possible to reshuffle the cards regarding the actors present and taking concrete action. First, the creation of Charentes Tourism (a tourism development agency common to the two departments) led to the disappearance of the Charente Departmental Tourism Committee, as well as the "Cognac Experience" system. The Poitou-Charentes Regional Tourism Committee also disappeared, together with "Destination Cognac-Vallée de la Charente". As collateral damage from the merger of departmental tourist institutions, "Les Étapes du cognac" association was dissolved in October 2019, weighed down by an economic deficit due to a decrease in public subsidies. However, the brand was taken over both by the BNIC and the Grand Cognac agglomeration. Nevertheless, the brand's future seems uncertain, given the desire to create a "Vignobles du cognac" brand as part of the application for a destination contract supported by Charentes tourism, the EPCI collective, the BNIC and the major trading houses.

To conclude, we can explain the strategy of the Cognac vineyard trading houses via the great autonomy of the houses (without the need for a cluster, nor for heritage recognition, nor for tourism development). In fact, their excellent economic health, ensured by sales worldwide and the notoriety of their luxury brands, justifies their low level of local involvement. Tourism professionals are well aware of this and still try to coordinate the supply of visiting circuits with the territory's tourist supply. However, common ground seems to be found in terms of a territorial *marketing* strategy, aiming not only to promote brands within the vineyard (the storytelling of production), but also to the territory's amenities (for attractiveness purposes).

1.4. Destination Anjou: a system developed by the CCI

This network was gradually formed and drew up its actions on a case-by-case basis, by seizing opportunities or in response to challenges. Its scope of action has always been within the perimeter of Anjou, a name chosen by local operators to counterbalance the low reputation of the region's institutional name, Maine-et-Loire, adopted in 1790 according to the canons of the time[16].

1.4.1. *Portrait of an association*

Destination Anjou (DA) is an associative structure which brings together the actors from different sectors in the Department of Maine-et-Loire: accommodation

16 As a matter of fact, in order to erase the names applied to the *Ancien Régime* territories, the Constituent Assembly decided to favor the terms resulting from the hydrographic network.

(25 hoteliers), catering (21) and leisure facilities (19). This case is characteristic for its original approach in relation to the CCIs (Chambers of Commerce and Industry), which generally focus on hotel clubs, or the coordination of a specific sector. In addition to existing sector organizations, others functioned within other sub-sectors, such as the Regional Federation of Tourist Sites (FREST). That federation was created in 2009 to lobby with public actors, particularly in relation to the financing which irrigates the sites supported by the institutions. Although that association no longer exists, its branch was taken by Activa, grouping 12 sites since 1994–1995 and managed by the CCI from 2007 to 2008. Other mergers between complementary actors can be mentioned, such as Escape Orne[17], but the latter is limited to marketing.

1.4.2. *Originally: a hotel club*

At the end of 1996, the original challenge for a group of six to eight professionals (brought together through mutual acquaintances) was to launch a quality approach as the one implemented in integrated chains. At the same time, there was a hotel club, but it was moribund. "We filled the void"[18]. Upon the arrival of Sylvain Dourneau, the CCI asked them what they could do differently, in comparison with a departmental tourism committee.

> The *Welcome to Anjou* initiative was a bit of a flop, because it aimed to bring together the actors, but without giving any precise specifications: it was open to whoever wanted in. At the chamber, we studied its feasibility, and we took off… from a hard core" (interview with Sylvain Dourneau).

Furthermore, the CCI seized this opportunity as a lever to bring together the three chambers of the department (based in Angers, Cholet and Le Mans), with a view to merging them under a single entity. In 1998, the statutes of an association were filed. The club worked on the quality approach. The initial catalyst was the need of independent hoteliers to resist competition from integrated chains. This was later reinforced by the chamber's organizational strategy.

"We started with a quality approach, because that is the mission of a CCI". The CCI welcomed the club and integrated it with others within the chamber's frame. Little by little, other missions were added, such as improving the quality of

17 Name of the tourist actor network formed by the Orne CCI to market its departmental offer.

18 Interview with Sylvain Dourneau, from the CCI, and Vincent Bouyer, Hôtel de France, and president of Destination Anjou. All quotes are drawn from this interview.

reception. Then, the association moved on to a more general action plan. From that moment on, the initiative brought together independent actors and chain members. The quality approach was abandoned in 2013, because it was no longer necessary: "everyone was doing it" and the France quality label was established.

> We didn't want to duplicate. The quality approach had two virtues... structuring the supply... and rallying the people... It also created a dynamic, a sense of competition, because we organized an event each time, for example, delivering the trophy for the best reception.

However, it was no longer essential. Afterwards, the emphasis was placed on other actions, such as communication and projects to enhance reception in Maine-et-Loire and value the territory's identity. An action plan was set in motion. A first website was created in 1998.

> The idea was to get the restaurateur out of his kitchen, the hotelier out of his room. And it was also a way to bring them together... The environment was modernized through the Internet tool, it functioned as a club, as a network to share good practices.

1.4.3. *Revitalization of the territory by a tourist actor*

In 2001, the association opened up to visiting sites. The action evolved towards a more global approach and the association strove to establish itself in the institutional landscape, by integrating different boards of directors (such as those from the Congress center, the Anjou Departmental Committee, Angers Tourism or Terra Botanica semi-public companies). The goal was to influence public policies. This development aroused opposition from the departmental tourism committee, which perceived this as head-on competition. A study ("Who does what in tourism?") made it possible to clarify the roles of each person and to highlight the complementarities between the two approaches. Similarly, when DA considered the implementation of a reservations system, the group backed down in the face of the identical initiative taken by the tourist office. A pragmatic approach was taken, with a focus on the coordination of networks. Initiatives were taken on an ad hoc basis. They were pursued, in case of success, and abandoned if the result was unsatisfactory.

Sometimes, the action had positive effects, but not necessarily the ones expected.

> For example, the privilege passport package. The idea was to sell the customer a privilege passport which enabled them to visit all the Destination Anjou sites, with a view to increasing the length of stays. We messed up a bit, we must have sold a hundred boxes.

Although it did not work, there were unexpected positive effects. This made it possible to increase internal dynamics by developing joint thinking, "to work in a slightly different way... to foster the learning side of working differently and of working together". In concrete terms, there was no shared action for marketing purposes, but the idea was to encourage "business referral". Although the creation of a reservations system was dismissed, an online availability platform was created, which enabled the actors to refer customers to another establishment in the event of unavailability.

Other opportunities were also seized.

> For example, the Cancer Institute asked us to organize a Michelin star banquet for which we all worked together. This increased our notoriety and created a new network within the network [...]. Beyond the marketing, communication, brochure printing operations, [it is] important for people to work together. Apart from the operations which specifically benefit members (an increase in the territory's notoriety), other events with a beneficial impact are organized, such as, for example "A dish for Anjou", a contest for the election of a typical dish of local gastronomy.

1.5. The Dordogne Valley: political ambition at the service of administrative division

The goal of local political actors was to propose a tourist destination consistent with tourist practices; whereby, erasing regional, departmental and intercommunal administrative boundaries. How has this ambitious political project been deployed and what are its limitations?

1.5.1. *Tourism in the Dordogne Valley*

The Dordogne Valley is an area that straddles the departments of Lot, Corrèze and Dordogne and whose boundaries are poorly identified. It is crossed by the Dordogne River, whose source is at Puy de Sancy, in Auvergne, and which gives its name to several territories it crosses until its confluence with the Garonne river. This area brings together the highest tourist numbers in the Lot department, comparable to the caves and valleys of Périgord (Milon 1960). In Périgord, the economic effect of tourism is marked by concentrated attendance along the lower Vézère valley and in the immediate surroundings of Sarlat (Delbru and Genty 1986). In 2018, this area (often referred to as "Périgord Noir" due to the color of the cork oaks which characterize the landscape) amounted to 59.09% of the accommodation supply, generating 42.78% of overnight stays[19]. Further upstream, in the Lot department, the bank of the Dordogne represented 34% of the department's capacity and accounted for 33% of overnight stays in 2018[20]. In addition, it is also along the Dordogne Valley that the most popular tourist sites are located. In the Périgord Noir, the town of Sarlat welcomes an average of 2 million visitors per year. For its part, the International Center of Parietal Art Montignac ("Lascaux IV") received 380,000 visitors in 2018, after welcoming 500,000 in 2017, when it opened its doors for the first time. Similarly, in the Department of the Lot, we can mention the village of Rocamadour, a known stopover for pilgrims on the way to Santiago de Compostela, which welcomes approximately 1.5 million visitors per year, or the *Gouffre de Padirac*, which received 480,000 visitors in 2018. In Corrèze, one village welcomes an outstanding number of tourists: Collonges-la-Rouge, with 700,000 annual visitors. The villages of Argentat or Beaulieu on the banks of the Dordogne also benefit from numerous tourist visits, thanks to their location within the same valley.

However, the geographical reality of the Dordogne Valley does not constitute an administrative reality. In 1974, the geographer Michel Daubet observed that there was an interest in promoting tourism in the "Dordogne Valley":

> Beyond this framework, tourist development must be considered on a regional level. Would there not be an advantage in actively developing a joint action to integrate the Vézère Valley, the curiosities of Haut-Quercy and the Tarn gorges, for example, under the same whole? Would it not be essential to encourage the tourist development of the Dordogne Valley throughout its middle course? (Daubet 1974, p. 316)

In 2017, Philippe Violier identified this perimeter as follows: high to very high intensity of attendance in the Périgord Noir, and medium to high intensity north of the Lot. The touristic phenomenon is expressed as a countryside divided between the distant outskirts of metropolises, where villages emerge from their torpor every weekend, and actual tourist destinations, frequented mainly during the summer. For tourist communications, Dordogne uses the historic appellation "Périgord" or "the Lot" (Violier 2017) in Stock (2017, p. 280). As it is customary, for their practices, they operate in combination (MIT team 2011). They are dominated by discovery, particularly those devoted to prehistoric heritage sites, which receive the greatest number of visitors. Castles, churches and medieval villages are abundantly present in the valleys. Many villages have also been awarded the "Most Beautiful Village of France" label (3 in the Lot, 2 in Corrèze, 9 in the Périgord Noir). Gastronomy has a

19 *Analyse Flux vision tourisme Dordogne*, Dordogne Departmental Tourism Committee, "Tableau de bord de suivi de l'économie touristique 2018", p. 7, p. 41.

20 *Analyse Flux vision tourisme*, Lot 2017, p. 26.

prominent place, through the promotion of local products. There are also entertainment activities such as canoeing down the rivers, for example, which offers a different perspective on the landscapes. Tourism is diffuse and manifests as "different non-concentrated forms of tourism which exist in non-touristy spaces, within versatile economies whose dynamics can be either declining or ascending" (Knafou 1995, p. 15). In concrete terms, the presence of tourism reflects an underlying tourist network, that is, the "networking of places through circuits, which reflect itinerant practices more or less influenced by voluntary actions" (MIT team 2005, p. 341). However, there is no tourist institution that embraces the entire Dordogne Valley. This is a result of the numerous administrative (departmental and regional) divisions. Tourism actors are also seeking to overcome these limitations by proposing new tourist organizations to cover the perimeter of the Dordogne Valley, including the Departments of Lot, Corrèze and Dordogne.

1.5.2. *Creation of a tourist office for the Dordogne Valley*

In 2010, the municipalities in the north of the Department of Lot gathered for developing tourism, the territory's main economic activity. Despite their opposite political tendencies, two elected officials agreed to bring together the tourist offices in the north of the Lot. These were Pierre Destic (right-wing mayor and president of the community of municipalities of Saint-Céré), and Gilles Liébus, (left-wing mayor of Meyronne and president of the community of municipalities of Cauvaldor). In 2009, the municipality of Rocamadour had already integrated the community of municipalities of Souillac, facilitating the creation of a Souillac-Rocamadour tourist office. In 2010, a joint tourist union was created in order to set up a new tourist office, under the management of the communities of municipalities of Souillac-Rocamadour and Padirac. In 2012, 55 municipalities spreading over two inter-municipalities came together and formed a first tourist office in the Dordogne Valley, administered by a mixed union. Then, 28 other municipalities joined them, convinced by the dynamics of the organization. The status of the tourist office was that of a public industrial and commercial establishment (EPIC, *Établissement public industriel et commercial*), supported by a new joint union. In total, eight communities of municipalities, seven tourist offices and nine permanent offices were merged.

The desire for a merger was driven by the will of pooling resources, something which motivated the elected officials. Until then, each of the seven tourist offices had managed their own brochures and websites, and the major service providers made their contributions to all those structures. The opinion *leaders* among the appointed officials and socio-professionals played a key role in encouraging other actors to follow the collective approach. Another challenge for the joint union was for each municipality to contribute to the tourist office's financing. In 2014, exchanges were established with the municipalities on the Corrèze bank of the Dordogne to integrate 65 municipalities and six communities of municipalities through the agency of the EPIC. The new tourist office of the Dordogne Valley was thus created on January 1, 2016. It included 148 municipalities located across nine communities of municipalities and adhered to two mixed unions (the Dordogne Valley region in the Department of Lot and the Corrézienne Dordogne Valley in the Department of Corrèze) (Figure 1.7).

1.5.3. *Aggregation of tourism actors from the Lot and Dordogne Valleys into an association*

In 2015, under the leadership of the CEO of the *Gouffre de Padirac*, a 50-member inter-industry departmental network was created. The association known as "Touristic Vallées Lot Dordogne" was born, thanks to the impetus of the Lot Chamber of Commerce and Industry and brought together different actors from the tourist sector: visiting sites, restaurateurs, accommodation providers, holiday villages, travel agencies, etc. From March 2016, meeting events were organized with the aim of creating "synergies between professionals". At the end of 2019, the association had 90 members, including one legal member, the Chamber of Commerce and Industry – which subsidized the association for the sum of €26,000 – eight active members (the Lot campsite association, the association of boat operators on the Lot river, the association of sites in the Lot and the Dordogne Valleys, the *Gouffre de Padirac*, the association *Bonnes tables du Lot*, the hoteliers association *Logis du Lot*, the *Site* Remarquable du Goût of the Lalbenque truffle market and the Lot Union of hotel trades and industries), as well as other adhering members whose annual subscription was set at €120.

Figure 1.7. *Schematic map of the territory of the Dordogne Valley Tourist Office (Piriou 2020b, p. 19)*

However, tourist institutions were conspicuous by their absence! For a good reason, the association was formed following the distrust of the Lot Departmental Tourism Committee. These professionals "wished to have their interests heard, to increase the territory's attractiveness and to gain a better understanding of the departmental tourist fabric"[21]. In fact, the idea was born during a business forum bringing together 70 persons in 2015, where tourism actors expressed the wish to create a new network. An institutional actor explained that the association's desire to sideline institutions and to create a network exclusively bringing together socio-professionals stemmed from the political opposition between the presidency of the Lot Chamber of Commerce and Industry and the Departmental Council of Lot. Confronted with economic difficulties, the Chambers of Commerce and Industry would seek to reaffirm their unifying role, to the detriment of other existing institutions. Anyway, the sites in the Dordogne Valley felt poorly represented by tourist communications from the Department of "Lot", because they designated the whole of the Lot Valley. In 2019, the Lot Departmental Tourism Committee reviewed this communication and created the destination brand "Vallées Lot&Dordogne". Now, was this the consequence of distrust or the product of reflection around a strategic positioning?

1.5.4. *Assertion of identity and political conflicts*

The territorial boundary of the Dordogne Valley Tourist Office finishes at the Department of Dordogne, and particularly excludes the Périgord Noir. Also, a partnership agreement has been established between the Sarlat-Périgord Noir Tourist Office and that of the Dordogne Valley, to ensure promotion among groups. Their respective websites also evoke their territories. The directors observed that tourists systematically came to Sarlat and Rocamadour, even though each of the tourist offices could not market the place located in the neighboring department. Furthermore, in 2015, a Dordogne Valley destination contract was signed for a period of three years, with multiple local authorities in Nouvelle-Aquitaine and Occitanie, including tourist institutions, tourist operators (such as the airports of Brive and Bergerac) and other socio-economic actors, under the leadership of the Public Territorial Establishment of the Dordogne Basin (EPIDOR, *Établissement public territorial du bassin de la Dordogne*).

Along the Dordogne Valley, to the north and to the east, both on the Lot side and on the Corrèze side, the process of bringing together the tourist offices marked the advent of a local institution strengthening the spirit of an interdepartmental tourist destination. However, difficulties persisted in the expansion of those areas, particularly towards Périgord Noir, located in the Department of Dordogne. Personalities played a key role in the extension of a territory of action. For the areas

21 Press release "Touristic Vallées Lot Dordogne", press conference from January 7, 2016.

located in Lot and Corrèze, despite being politically opposed, the two elected officials agreed on the challenges of tourism development. Many appointed officials, not to say almost all of them, governed the tourist offices. In the Dordogne Valley, the political will was to achieve a merger between the Corrèze tourist offices, as evidenced by the comments of departmental and regional elected officials reported in the regional daily press in March 2013:

> Christelle Coursat, regional councilor, president of the Limousin Regional Tourism Committee, first mentioned the common base of tourist offices and the Region's support policy, which made it possible to better structure, federate and pool resources. [...] Jacques Descargues, vice-president of the General Council [of Corrèze], congratulated those responsible for having anticipated the merger: he declared that the merger had long been underway in the department. It is more complicated for some, nowadays everyone has identified the need to move towards the tourism of the future[22].

Furthermore, the tourist office was not limited to a financing or sponsorship role, as evidenced by the involvement in destination contracts. These conventions allowed tourism actors to become engaged in the governance of the destination, enhancing its attractiveness and competitiveness (Bédé 2015). The dynamic of cooperative actions brought to fruition by the directors of the tourist offices of the Dordogne Valley and Sarlat-Périgord Noir resulted in the signing of a destination contract. This took place between the EPIDOR and other co-signatories: the state, three regions, two airports, six departments, EDF and four tourist offices. However, unlike the Loire Valley, there was no consensus around the Dordogne brand, limiting many activities to the Dordogne departmental boundaries, even though the project called for their loosening.

1.5.5. *Ensuring equal governance between Lot and Corrèze*

In the Dordogne Valley, as soon as the Lot tourist offices were brought together, the status of a public industrial and commercial establishment was chosen, with a president at its head (who was also to preside over the supervisory Lot community of municipalities), and a director. This was a status created by deliberation of the local supervisory authority, which acted as a legal entity under public law with financial autonomy. However, unlike an association, interventions were limited to local authorities' territories. This status implied a strong political orientation, something which was felt all the way to the management committee, since it mainly comprised elected officials. Within the "Dordogne Valley" association, the

22 Excerpt from the article "Tourisme : la fusion des offices de tourisme à la traîne". *La Montagne*, March 15, 2013, p. 17.

management committee of the tourist office was made up of 15 elected officials from the Department of Lot and 12 appointed officials from the Department of Corrèze. Nevertheless, the originality of governance was particularly striking due to the parity between Lot and Corrèze, mainly through the support of the same number of 10 socio-professional representatives for each of the two departments. To elect this college of socio-professionals, public meetings were organized, bringing together 10 sectors identified with the tourism sector, and during which the elections of representatives took place. The communities then only had to appoint the elected persons. This approach – previously implemented when the tourist office was only set within the Department of Lot – signaled an innovative mode of governance, because at the time, it already included 11 elected officials and 10 socio-professionals. Then, the tourist office benefited from a budget of approximately 2 million euros, including €600,000 from the local supervisory authorities (which provided €10.50 per inhabitant), as well as assistance for a total of €70,000, contributed by the Region of Occitanie and €40,000 by the Nouvelle-Aquitaine region.

The former directors of the seven tourist offices after the merger were entrusted with executive missions within the new structure. Within the Dordogne Valley Tourist Office, the organization chart evolves every year to come out of "silo" mode and to privilege a project management modality. For example, two employees are project leaders, and two other employees are project managers. Due to this, staff training seems necessary to standardize skills. In the Dordogne Valley, the Corrèze bank and the Lot bank did not function in the same way. It was necessary to find solutions to share documents among 148 municipalities whose premises were over an hour's drive away. Staff were trained in the use of clouds or even a telephone system facilitating rapid exchanges between sites. A person was also dedicated to managing calls, with switchover scenarios, depending on office staffing levels. However, in both tourist offices, particular importance was given to reception teams (with continuous openings during the high season), as tourist activity is strongly related to the notion of service.

1.6. Tourism in the Lake Geneva region: the challenge for actors to step beyond boundaries

Léman Sans Frontière is an association which in 2023 brought together 29 leisure and cultural sites located in France and Switzerland on either side of Lake Léman, also known as Lake Geneva. Their unifying factor is the lake, which has catalyzed their image as part of a marketing collaboration, which has now lasted for almost 30 years. Whether they be interpretation centers, local transport companies, amusement parks, castles, stations, etc., in France or in Switzerland, all the member sites of the association coordinate this voluntary network on behalf of their organization, in view of creating cross-border tourist destination for the collective good.

Communication with the public represents one of the main pillars of the association's activities: in 1996, a first guide was published, bringing together partner sites. This guide is distributed in each of the sites as well as in various places of passage in the Lake Geneva region (accommodation, restaurants, train stations or airports, tourist offices, etc.). During the 2000s, the sites' information was digitalized, thus enabling the sharing of an unprecedented Franco-Swiss database. Since the beginning of the 2010s, a new direction has been given to the creation of a cross-border tourist supply, including visiting sites, accommodation and transport.

1.6.1. *The Lake Geneva basin, tourism development on the French and Swiss shores*

The Lake Geneva region is an area which extends around Lake Geneva. This region, which straddles between France and Switzerland, has two cities which are known worldwide. Located along the boundary with France, the Swiss city of Geneva is the largest in the region in terms of the number of inhabitants. It has 185,791 inhabitants, to which must be added the 208 municipalities forming a cross-border agglomeration of nearly a million inhabitants[23], Greater Geneva. From the 18th century, the city positioned itself as a literary crossroads, justifying the arrival of artists. Jean-Jacques Rousseau bore witness to the beauty of the lake landscape:

> The view of the Lake of Geneva and its lovely shores had always a particular attraction in my eyes, which I cannot explain, and which does not depend only on the beauty of the sight, but on something more compelling which moves and stimulates me (Jean-Jacques Rousseau, *The Confessions*, Book IV, 1953[24]).

The city also stands at an important geopolitical crossroads thanks to the presence of international institutions such as the League of Nations or the United Nations since the 20th century (Tissot 2007).

Fifty kilometers away on the northern shore of Lake Geneva, there is the city of Lausanne, with 139,720 inhabitants. Since the middle of the 19th century, this city

---

23 Sources: *Mémento statistique du canton de Genève 2018*, Cantonal Statistical Office, Geneva, June 2018, p. 2 and *Grand Genève, mode d'emploi*, p. 2.

24 Excerpt from the pedagogical dossier of the exhibition "D'une rive à l'autre, voyage autour du Léman, acte II" from March 24 to November 10, 2018, town of Thonon-les-Bains, Department of Culture and Heritage, p. 14.

has benefited from the connection to an international transport network and its educational, sporting and healthcare establishments have an unparalleled reputation. Even today, the International Olympic Committee is set there (Humair 2007).

The region has benefited from increased tourism through the development of several resorts along the lake's shores. At the beginning of the 19th century, Montreux, located at the eastern end of the lake, was no more than a group of villages. The municipality has become a real fashionable resort, with a densification of high-standing hotel establishments (Rinaldi 2006). Together with the town of Vevey (19,004 inhabitants)[25], Montreux (26,006 inhabitants) forms the "Vaudoise Riviera", a name definitively describing the region by the tourist industry.

On the French shore, other locations have developed to treat patients: Thonon-les-Bains, which has 34,895 inhabitants, and its neighbor less than 10 kilometers away, Évian-les-Bains, 8,965 inhabitants[26].

Évian-les-Bains experienced its real tourist development after the discovery of the Cachat spring in 1789, spurring the creation of the famous brand of mineral water with the same name, and the establishment of a thermal spa resort in 1826. The thermal station development was such a success that the bourgeois resort model was replicated around the small outlying towns over the lake's banks (Maxilly, Publier, Neuvecelle). At present, Évian-les-Bains is positioned as a high-end resort, competing with Lausanne, located on the opposite bank (Sceau 1974; Durand and Fererol 2009).

The Lake Geneva region includes:

– over 3 million inhabitants[27], including the French and the Swiss side of Lake Geneva;

– tourist and business attendance estimated at over 8.5 million overnight stays in 2017[28];

– 28,576 establishments involved in the tourism sector, including 37% catering, 31% leisure activities, 15% accommodation and 10% transport[29].

25 Source: Statistical office of the Canton of Vaud, Vaud population as of December 31, 2018.

26 According to INSEE in 2015.

27 Source: *Conseil du Léman*, http://www.conseilduleman.org.

28 Sources: 3,054,435 nights in Geneva according to the Geneva Cantonal Statistics Office (2017), "Bilan de l'hôtellerie genevoise", *Informations statistiques*, No. 11, June 2018, p. 2; 3,467,900 overnight stays in Lake Geneva, Vallée Verte, Les Brasses, according to Savoie Mont-Blanc Tourism (2017), "Chiffres clés du tourisme en Savoie Mont-Blanc", p. 9; 1,227,800 overnight stays in Lausanne according to the tourist office of the Canton of Vaud (2018), "Nuitées 2017: une nouvelle année record pour le tourisme", press release from the tourist office of the Canton of Vaud, February 22, 2018.

However, there is no official tourist destination that encompasses the whole Lake Geneva region. This can be attributed to the multiple borders within this region marking territorial boundaries. This situation complicates the relationships between actors on either side of the frontiers (Piriou 2011).

The lake separates two nations: France and Switzerland. Then, within each region, there are other subdivisions. The Swiss part is divided into three cantons (Geneva, Vaud and Valais), while the French part has been split into two departments (Ain and Haute-Savoie). Moreover, these territories are further divided into inter-municipalities and municipalities in the French sector, and into districts and municipalities in the Swiss sector.

Until the 1970s, cross-border relations between France and Switzerland were faint-hearted (Poschet et al. 2001). However, in 1987, the creation of the Lake Geneva Council, a political body for cooperation between the French departments and the Swiss cantons of the Lake Geneva region, initiated collective reflections, particularly for facilitating a cross-border space (banking, insurance, transport, etc.) for a regional population, namely those who cross the national boundary daily as part of a home–work commute. On the contrary, when it comes to tourism promotion, each territory sells its administrative perimeter. In Geneva, for example, the Geneva Tourism and Congress Foundation (which serves as the tourist office), together with the Tourism Advisory Commission, are responsible for the application of the cantonal law on tourism:

1) The purpose of this law is to ease the promotion and development of tourism.

2) In concrete, it aims a) to develop quality tourism matching the demand, highlighting natural and historical riches, cultural and sporting events as well as the traditions of the canton of Geneva; b) to stimulate the promotion of tourism for Geneva; c) to support the economy through the development of tourism (Cantonal Law on tourism in Geneva, Article 1).

The promotion of the tourist destination and the organization of events are limited to the city and canton of Geneva. This same mechanism applies to all the other territories in the Lake Geneva region (Piriou 2018a). The cross-border character of this space (which reflects an idea of passage, of crossing) does not seem to be evident. In 2002, on this subject, referring to the case of the cross-border region of Lille, Pierre Mauroy[30] declared:

---

29 Source: 2014 figures according to the *Union lémanique des chambres de commerce* (2016), "Filière tourisme. Le bassin lémanique: une réalité économique", 2016 edition, p. 4.

30 Political figure in the city and metropolis of Lille from 1967 to 2013 and former French Prime Minister from 1981 to 1984.

> Cross-border organization is not an obligatory thing; it requires the will of everyone on both sides of the boundary (MOT 2002, p. 15).

Several French and Swiss leisure and cultural sites, private owners, and then public actors have come together since 1995 to collaborate in the marketing of a cross-border tourist destination.

1.6.2. *Genesis of a Franco-Swiss tourist association*

It was in 1994 that the idea of creating Léman Sans Frontière was born, following the initiative of actors from two distinct territories: the pre-Alpine resort of Moléson-sur-Gruyères and the lakeside municipality of Bouveret, represented by their promotion corporations (tourist offices). Established in 1995, the association aims to promote Franco-Swiss tourism, so as to increase tourist visits around Lake Geneva. For the small Moléson resort, designed in the 1960s by Philippe Micheloud, the aim was to capture summer tourist flows. In fact, this private management station is located at an altitude of 1,100 meters. However, the limited number of days of snow cover quickly prompted developers to reorient the resort's seasonality, by betting on summer activities, which bring in between 60% and 80% of the annual turnover. For the small lakeside municipality, there was a desire to convince vacationers who go to the mountains during the winter to make the most of the lakeshore during the summer. Under the leadership of Charles-Henri Coutaz, president of the promotion corporation (tourist office) of Bouveret, the project was to intensify tourist attendance in the municipality around the Swiss Vapeur Parc (created in 1989, a park dedicated to locomotives and miniature railways).

1.6.2.1. *Sharing clientele and making tourists sightsee around the lake*

This principle of having visitors circulate around the lake then arouses the interest of other actors, such as the *Compagnie Générale de Navigation* (a navigation company over Lake Geneva), or the Evian mineral water bottling company. With 12 sites in 1995, 16 sites in 1996 and 28 in 1999, the association's number of members rose to 26 in 2023 (including 24 sites in Switzerland and nine sites in France). Among the association's members, there are local organizations (most of which are private, such as tourist train sites, gardens, ski lifts, etc.). These local actors also have a say in public and parapublic tourism institutions, with more of a development (or even a prescribing) role. They are mainly tourist offices (in France: Yvoire, Divonne-les-Bains, Monts-Jura, Thonon-les-Bains, etc.; in Switzerland: Romont, Gruyères).

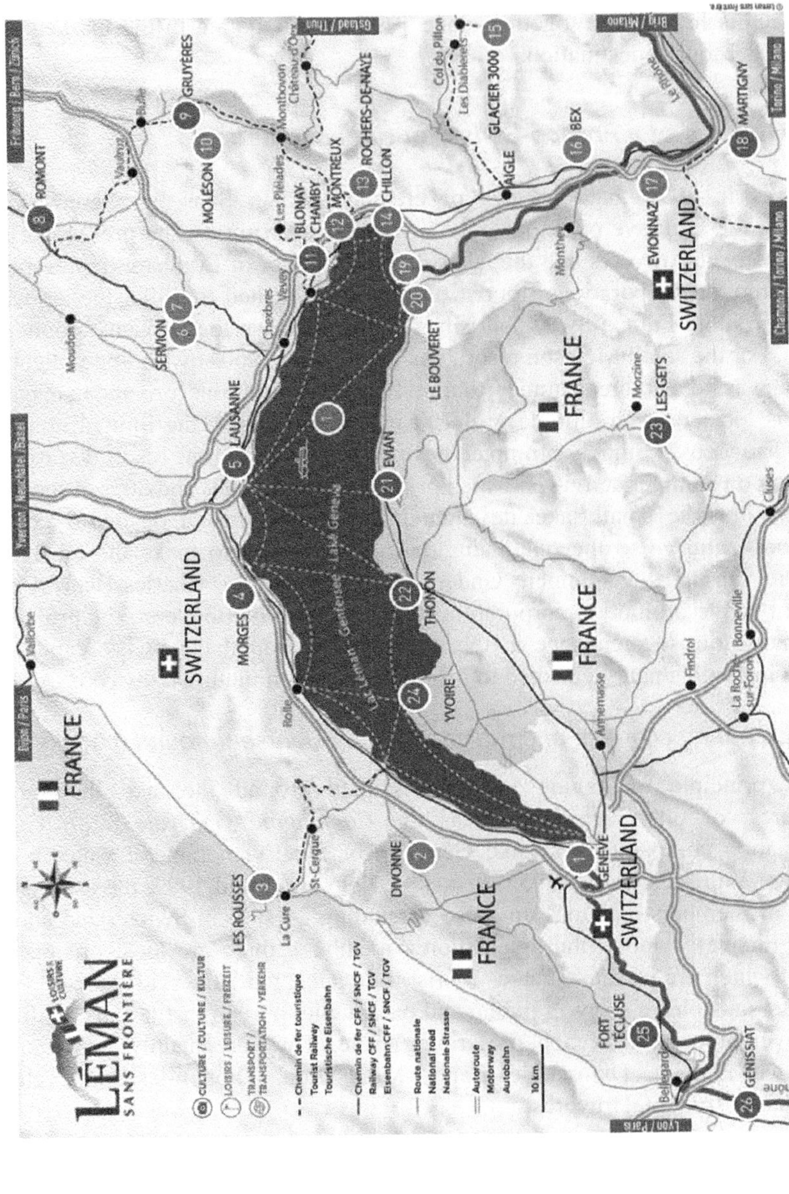

Figure 1.8. *Map of the 26 sites of the “Léman Sans Frontière” association in 2023 (Léman Sans Frontière, 2023)*

COMMENT ON FIGURE 1.8.– *Among the members*[31]*, some sites have significant attendance: 365,500 visitors to the Chillon Castle, located near Montreux (Switzerland), 304,800 visitors to the Olympic Museum in Lausanne (Switzerland); others have more modest attendance: 36,869 visitors to the Jardin des Cinq Sens in Yvoire (France); 24,080 visitors to the Château de Ripaille in Thonon-les-Bains (France).*

All these actors are associated in this Franco-Swiss tourism development project. According to the statutes of the association, the starting point is:

> To bring together tourist attraction sites with significant visitor potential and supply diversification so as to boost promotional forces and better satisfy the audience, providing them with a variety of quality excursions[32].

A directory guide was also developed, with the purpose of reaching out to tourists. Every year, 275,000 copies are published and distributed to all member sites and to the various information and promotion points (accommodation, catering, tourist offices, etc.).

1.6.3. *Defining a common strategy for French and Swiss actors*

To build a strong presence in the landscape of French and Swiss tourism actors, the Léman Sans Frontière association must define a strategic axis. As an association, the difficulty for Léman Sans Frontière is to bear in mind collective interest, all the while respecting the individual interest of each member. The issues at stake are not the same for all the members:

– A water park, such as Aquaparc, located on the Swiss bank of Bouveret, welcomes over 260,000 visitors per year and generates a turnover of over nine million Swiss francs[33].

– A museum, such as Voltaire's castle in Ferney (France), now a public property, welcomes 13,924 visitors and collects €1,962 in entry fees over a year[34].

31 According to 2016 figures communicated by the heads of tourist offices and tourist sites following field surveys (Piriou 2019).

32 According to Article 2 of the 1995 statutes of the "Léman Sans Frontière" association.

33 Anonymous (2013), "Aquaparc, en route vers quatre millions de visiteurs", *Le Temps*, June 10, 2013, https://www.letemps.ch/economie/aquaparc-route-vers-quatre-millions-visiteurs.

34 *Centre national des monuments historiques* (2017), "Rapport d'activités 2016", p. 67.

Still, all the sites have contributed 7,000 Swiss francs (approximately €6,400) to the Léman Sans Frontière association over a period of three years. For some, this sum only constitutes an item within the communication budget; whereas, for others, it may represent a real complex financial arrangement, with multiple actors involved and negotiated to the nearest cent. Despite this, the positive aspects of an alliance and the network strategy prevail, embodying a valuable opportunity for member sites. This strategy reinforces commercial visibility and grants access to an enhanced distribution capacity (Frochot and Legohérel 2018, p. 156). Resource pooling makes it possible to achieve significant savings.

The association brings together sites with a prominent international reputation, thus attracting significant tourist flows. Furthermore, these sites act as locomotives for tourist flows to be spread towards smaller sites. They also benefit from the structure's communication (as in the case of the Olympic Museum in Lausanne, or the Evian mineral water bottling plant).

The number of member sites (26 sites in 2023) reveals that the association has branches for managing tourist information, both in terms of dissemination and collection. Whether located in France or Switzerland, regardless of the department or canton, each member distributes the tourist guide, all the while collecting data as to the locality's policy or its tourist attendance. This valuable deployment makes it possible to overcome the constraints of boundaries when it comes to obtaining of statistical data or disseminating communication media. For example, some sites establish partnerships with brochure distribution companies.

At every membership renewal date (every three years), some member sites leave the network, mainly due to financial reasons or changes in management. Except for 15 member sites, which joined the Franco-Swiss collaboration between 1996 and 1998 (and which were still present in 2023), around 10 sites left the network a few years after their first membership. This instability makes the alliance and network strategy envisaged by member sites difficult to pursue.

However, currency fluctuations have an undeniable impact on tourist traffic. For example, in 2015, the Swiss franc was revalued (appreciation of over 30% during the same week). The purchasing power of French tourists in Switzerland is rendered more difficult by the presence of a strong franc. As the Swiss Confederation is outside the European Union, this monetary boundary also complicates alliance and network strategies.

1.6.4. *Maintaining the Franco-Swiss collaboration over time*

1.6.4.1. *Shared cross-border collaboration operations*

To encourage the joint promotion of these combined service initiatives, the Léman Sans Frontière association financially supports member sites[35]:

– 500 Swiss francs (approximately €445) per partner in the context of general public exhibitions (e.g. Mahana in Lyon and the International Tourism and Travelling Exhibition in Colmar);

– 500 Swiss francs per salon in the context of professional exhibitions (e.g. Top Résa in Paris or exhibitions for works councils in Dijon, Nancy, Annecy, etc.).

These salons are located in two well-identified trade areas, bordering the Lake Geneva region (eastern France, Rhône-Alpes, etc.). The construction of a supply including several sites reveals an underlying desire to encourage greater circulation of visitors in the Lake Geneva region, and mainly to increase the attendance of the association's member sites.

Striving to take a leap ahead in the process of combining supplies, the Léman Sans Frontière association requested financial assistance from the Interreg IV France/Switzerland program (2007–2013). A tandem that brought together the tourist office of Divonne-les-Bains (France) and the Swiss Vapeur Parc – both implementing the project on behalf of the association – made it possible to obtain funding for an amount of €556,250 (890,000 Swiss francs), thus marketing 80 supplies in the form of excursions or stays with a total cost of €943,750 (1.51 million Swiss francs).

In 2012, a reflection was initiated in order to reach other visitors, namely local population and tourists who do not consume the combined services, but who may be led to Mix&Match member sites at their will, as part of a discovery trip for leisure and cultural activities[36]. Without a doubt, at the beginning of the 2010s, the digital tool seemed undeniably useful to visitors for disseminating tourist information.

Between 2013 and 2015, the tourism development agency from Ain (Ain Tourisme, France) and the Bex Salt Mines Foundation (Switzerland) both carried out a Lake Geneva itinerant project called "Mobi- Léman", on behalf of the Léman Sans Frontière association and the Léman Council. Total project cost: €537,916.

35 Figures communicated to the general assembly of Léman Sans Frontière on April 20, 2007 in Servion.

36 Minutes of the general assembly of Léman Sans Frontière of October 3, 2012 in Neydens, p. 3.

€148,500 in European grants as part of the Interreg IVA France/Switzerland program.

At the operational level, the "Mobi-Léman" project was organized as follows:

– Working groups in each territory (departments and cantons) were made up of representatives from the spheres of tourism and culture.

– Each working group proposed three itineraries in 15 points of interest and wrote scenarios from a documentary base and using technical means (videos, augmented reality, etc.).

– 19 thematic itineraries were created, on either side of the shores of the lake, with a geolocation and coordination service offered via a smartphone application.

– Links between member sites via routes using different modes of transport (car, bike, boat, on foot).

At the end of 2016, the application already had 3,140 downloads.

1.6.4.2. *Developing synergies between private and public actors*

On April 12, 2016, the 20th anniversary of the Léman Sans Frontière association was celebrated. About a hundred guests gathered on the boat *Henry Dunant* (Compagnie Générale de Navigation fleet), during a festive cruise on Lake Geneva between the ports of Lausanne-Ouchy (Switzerland) and Évian-les-Bains (France). The president of the Lake Geneva Council in 2016 and Ms. Virginie Duby-Muller (MP for Haute-Savoie and vice-president of the Departmental Council of Haute-Savoie) testified to the role of the association in the relations between the two shores of Lake Geneva:

> It is true that, through the numerous cross-border projects developed since the beginning of the 2000s, Léman Sans Frontière illustrates, the Franco-Swiss tourism initiatives implemented in recent years, in a concrete and exemplary manner.

The recognition by public institutions of the concrete character and exemplary nature of a private initiative to promote exchanges between France and Switzerland confirms the success of Léman Sans Frontière. The alternating presidency of the Franco-Swiss association has contributed to establishing trust among members for 20 years. As stated by the current president, Mr. Philippe Benoît, during the anniversary ceremony, the challenge for the association is now to direct activities on two registers:

> To increase the notoriety of the association to the point of making Léman Sans Frontière *THE* essential network of Franco-Swiss sites[37].

To be an essential network, the association must ensure the solidity of its operation. With this goal in mind, the extraordinary general meeting of April 22, 2015 validated the clarification of the association's statutes. Two elements were to strengthen the collaborative mindset as well as the network's functioning.

Firstly, in the goals of the association, the new statutes of 2015 insisted not only on localization, but also on sharing (Article 3):

> To bring together operators from cultural, leisure or transport activities related to tourism in the Lake Geneva region in France and Switzerland. To develop and promote the activities of its members through mutual exchanges and joint actions.

Secondly, the organization of the association is defined by three distinct bodies (Article 5): the general assembly (supreme body, according to Article 6); the committee (appointed in a general assembly with at least five members, including two per region, an office comprising a president, a vice-president, a treasurer and another member elected, according to Article 17); and the audit body (responsible for verifying the accounts, according to Article 22).

The desire to promote the work of the committee and the office was particularly emphasized, from 2014, under the French presidency of Mr. Yves d'Yvoire (owner of the Jardin des Cinq Sens), when the dozens of underlying meetings were synthesized behind the motto: "Léman Sans Frontière, an association that works"[38]. It was repeated in 2017 under the Swiss presidency of Mr. Philippe Benoît, director of the Bex salt mines, who summarized the work of the year 2016, marked by the 20th anniversary[39]:

> 2016 was also a lot of work, 2 traditional general assemblies, 6 committees, 4 educational days with more than a hundred participants, around twenty trade fairs in Switzerland and France with several thousand contacts, 250,000 brochures distributed in 3 languages.

---

37 Léman Sans Frontière (2016), "Vingt ans de Léman Sans Frontière: une véritable *sucess story* franco-suisse", press release, April 12, 2016, p. 2.

38 Minutes of the general assembly of Léman Sans Frontière of Wednesday, April 9, 2014 in Lausanne, p. 3.

39 Minutes of the general assembly of Léman Sans Frontière of April 27, 2017 in Divonne-les-Bains, p. 3.

Léman Sans Frontière is an association that operates on a budget of 280,000 Swiss francs (€186,000), without employees. Besides, 80% of this sum is dedicated to carrying out promotional and communication actions. However, beyond promotion and communication, the association strives to increase cooperation between members. Cooperation also conveys the idea of exchange between members. It can be analyzed by the dialogue set in motion as well as by the role occupied by the actors in this same cooperation (de Bruyn and Fernández 2012). Also, several general assemblies were the subject of thematic conferences for members, depending on the concerns of the moment (Table 1.1).

Date and place	Theme
May 11, 2016 at Fort l'Écluse (France)	Tourism jurisdiction and French law of August 7, 2015 relating to the new territorial organization of the Republic.
April 27, 2017 at Divonne-les-Bains (France)	Symbiosis and importance of links between the tourist office and socio-professional in promotion and marketing.
May 4, 2018 at Gets (France)	Shedding light on cross-border tourism collaboration in the Portes du Soleil area.
November 30, 2018 at Prangins (Switzerland)	Knowledge of German tourism and expectations of "group" customers.

Table 1.1. *Examples of thematic conferences given during general assemblies of Léman Sans Frontière (2016–2018)*

To conclude, the approach launched by private tourism actors almost 30 years ago is part of the logic of cross-border tourism development in the administrative territories of the Lake Geneva basin. In addition to cooperation for the purposes of promotion and marketing of tourist sites and places, the search for a mechanism for understanding the functioning of partners (French and Swiss legal systems, setting up projects with European funding, etc.) has intensified relationships between the actors. A hint of fragility is nonetheless noted, given the fact that certain personalities contribute to maintaining exchanges and initiating collective projects. However, once these personalities leave the network, the intensity of exchanges may receive the impact of the difficulty for successors to become aware of how the network works, or a limitation of commitment.

1.7. *Autour du Louvre-Lens*: a UFO in a disadvantaged economic landscape

1.7.1. *An emerging tourism project*

1.7.1.1. *History*

Unlike other destination contracts aiming to draw attention towards tourist regions already acknowledged as such, the *Autour du Louvre-Lens* (ALL or "Around the Louvre-Lens") initiative was born to support an emerging tourism project. In any case, this is what happened in the territory where a new prestigious museum (the Louvre-Lens) was established, in order to enhance cultural appeal and thus regenerate a region subject to major economic and socio-political challenges.

After hemming and hawing over its location, the Louvre-Lens Museum was built in Hauts-de-France (then called Nord-Pas-de-Calais), a former mining territory, at the heart of a triple-contour space:

– the closest area (the district of Lens-Liévin, likely to obtain direct effects from the museum's implantation);

– the extended area, reaching the mining basin and its immediate fringes (from Béthune and Bruay-en-Artois to Carvin and Hénin, Douai and Valenciennes);

– on a regional scale, taking into account the attractions of other important cities such as Arras (to the south-east) or Lille (to the north). In a broader sense, the project aimed to stimulate the coastline of Nord-Pas-de-Calais and the Somme, the Flemish Region and the major tourist towns of neighboring Belgium.

Historically, the idea of setting up the Louvre at Lens was supported by politicians, in particular, by Mr. Daniel Percheron (the regional president at the time). Without his contribution, nothing would have been achieved. It was later supported to varying degrees, by the elected officials of the territory in question, not without local reluctance or conflicts. Special emphasis was placed on other complementary issues at stake: integrating popular cultures, acknowledging employment needs in the broad sense and considering other tourism and culture vectors (e.g. sport, ecology and green spaces, the conviviality of well-living and a festive atmosphere). The Region was a seriously involved partner in the project's financing and steering.

1.7.1.2. *An incipient tourist orientation*

Although the Louvre-Lens did not in itself awaken an incipient tourist function, it undoubtedly stimulated it, through the political attention it received, and the financial resources that were put into place in complementary goals (economic recovery measures, handcraft, and industrial clusters, etc.).

To boost tourist activity, it simultaneously benefited from four complementary attractiveness factors (beyond many other surrounding museums and cities already mentioned, such as Arras, Douai or Lille):

– the increasing appeal of places of remembrance (international cemeteries, wartime memories, notably the IWW on the occasion of its centenary, widely celebrated in France[40]);

– the valorization of the mining basin, which was included on the World Heritage List around the same time[41], through long-term work carried out locally. This declaration was accompanied by the restoration of several major sites and the development of visits to slag heaps or mining museums;

– the desire to develop circuits and green spaces and to connect them together, for walking, cycling or relaxation, in a territory where ecology had long been neglected in favor of landscape-destroying industrialization;

– the prestige of the urban heritage of certain cities, the belfries[42] (a regional feature crossing the Belgian border) and the art of celebration (carnivals, the Processional Giants[43], clearance sales, fun or sporting events) to awaken an original popular culture.

1.7.1.3. *Tourism as a way out of the economic crisis*

The original idea of the regional president was to rely on examples of regenerative ecosystems from implantations with tourist appeal and development models related to culture, services and the ecological transition.

As usual, it drew on an entanglement of models from different places: for example, Bilbao's revitalization thanks to its dynamic heritage, the relocation of

40 A pass bringing together 11 museums was also designed to create a sort of network of possibilities for visitors, the most popular being Ypres (in Belgium), Péronne (Museum of the Great War), Albert (Somme Museum 1916) and Arras (the Wellington Quarry). However, near Lens, we find the great symbolic places of Notre-Dame-de-Lorette (one of the largest soldiers' cemeteries) and Vimy (with its Canadian memorial). To a larger extent, there are around 50 memorial sites (memorials, cemeteries, museums) in the region, including in Lens itself (Lens 1914–18). According to C. Douay (Somme Museum): "before the start of the commemorations of the Great War, 1914–18 stood for Verdun. Today, more and more French citizens are coming to discover our places of remembrance, even if the British still make up the bulk of visitors. Among the British, almost every family has had one of its members fall on the Northern Front".

41 June 30, 2012.

42 Let us mention those of Douai (14th century), Béthune (14th century) and Arras (15th century), which echo many others in Flanders.

43 *Les Gayants de Douai* are particularly well known.

Beaubourg to Metz, the simultaneous creation of Mucem in Marseille, the Ruhr's older effort to transform its economy and landscape after being Germany's industrial and coal center.

To get out of the economic depression and the stagnation of its residents, it aimed to create jobs, relaunching activity in local towns, particularly in Lens (through Euralens); clusters in the handicraft and new technology fields were thus foreshadowed.

This spin-off policy, relying on the previously mentioned tourist sources, proved to be complex. It sometimes put the cart before the horse. It experienced delays (due to lack of operational branches and resource distribution, the implementation being more complicated than expected, as a result of conflicts between the actors involved). However, contrary to what some hasty analysts may have thought, the conflicts did not extinguish the movement, nor did they reduce the Louvre-Lens to the dimension of a heritage niche with weakened resonance.

Figure 1.9. *Example of mining rehabilitation, following inclusion on the World Heritage List, Wallers-Arenberg site, in Valenciennois. Picture by Pierre Louart, 2019*

They rather created a coopetition atmosphere between local towns, with the emergence of other projects. This happened not only between culture and tourism actors who had learned how to work a little better together, no matter what, but also between the culture imported to the Louvre-Lens and local traditions (which resurfaced by expressing their own demands). In short, the territory found itself in a situation of increasing coopetition.

1.7.2. *The logic of a regenerative territory*

1.7.2.1. *Mining heritage*

Even though we can associate the Louvre-Lens with other cultural or tourist markings in the region, its focal point is the mining basin[44]. The testimonies from the promoters, as well as from the park's director and curator of the main collections place a special emphasis on the historical roots of the place as a coal mining site. Moreover, the architects of the SANAA agency remodeled their initial project at the last minute to harmonize it with the environment ("our intention was for the visitor not to note where the landscape ends and the building begins").

Despite an initial emphasis on the metropolis of Lens-Liévin, the cultural impact was necessarily broader, thus extending the tourist scope to the mining basin area, opening up to the region at large. It is in this space that the museum's circuits and resonances were gradually organized.

In fact, the entire mining area experienced unemployment challenges (generally higher than elsewhere in Nord-Pas-de-Calais), and difficulties recovering from its industrial past. In that sense, the territory was relatively homogeneous, compelled to envision its renewal[45].

44 "At present, the former mining basin constitutes a dense and uninterrupted space of urban municipalities making up the densely populated districts of Béthune, Lens, Douai, and Valenciennes, with respectively 285,000, 362,000, 248,000 and 349,000 inhabitants. The impact of the Louvre's arrival to the heart of a former mining basin will go far beyond the sole territory of the municipality of Lens. Due to the high intensity of transport and communication infrastructure, the city of Lens is connected to the main economic development centers, at the heart of Lille's metropolitan area, which extends from Béthune to Valenciennes, from Arras to Lille and beyond the Belgian border. This calls into question the pertinence of the territorial scale when engaging public policies for supporting the implantation of the Louvre", *La Voix du Nord*, 2012.

45 Facing problems not very different from the ones experienced by the Ruhr in Germany, without completely succeeding in overcoming difficulties.

It could be interpreted as a potentially regenerative[46] territory, which would likely have its image revitalized by tourism and encouraged to innovate in order to attain socio-economic rebound.

To contribute to the territory's economic rebirth and its social rehabilitation, it was necessary to envisage new equipment, new projects, new deployments of services or industry, as well as work on town planning, culture and overall symbolism.

1.7.2.2. *Chronology*

In order to attain those goals, let us cite some key dates in five points:

– The preparations were like a stroke of luck converted into necessity. Special praise should be given to Mr. Daniel Percheron, president of the Nord-Pas-de-Calais region (Liévin) who took advantage of his connections with the former president of the Louvre (H. Loyrette). He showed a lot of persistence to see the project through. In 2003, Guy Delcourt, mayor of Lens, had already proposed the city's candidacy to host the decentralization of a large Parisian cultural establishment (initiative of the then Minister of Culture). Lens was chosen for the Louvre implantation in 2004. The laying of the first stone took place on December 4, 2009.

– At the same time, Euralens was formed (2009), leading economic, urban and tourist projects (e.g. *La chaîne des parcs*): "born with the arrival of the Louvre in Lens, Euralens is above all a major forum for the territories, a collaborative approach for accelerating the urban, economic, social and cultural transformation of a metropolis in the making" (www.euralens.org). At the same time, within the framework of local development, there were more or less concerted groups (museums, slag heaps, regional towns, tourism, etc.).

"On 21 December 1990, the last *gaillette* (a large piece of coal) came up from pit No. 9/9 bis in Oignies. Around twenty years later, the territory of the former mining basin, still marked by the industrial revolution and past coal mining activity, saw the establishment of the Louvre in Lens and had to prepare to accommodate the 500,000 visitors expected every year under the best possible conditions. Public actors supported the economic reconversion and the numerous transitions that this territory went through, classified by UNESCO in 2012 as an evolving cultural landscape. As part of the Regional Economic Development Plan (SRDE, *Schéma régional de développement économique*), the Nord-Pas-de-Calais region launched Euralens, a tool for the transformation of the territory and a unifying structure for projects".

Box 1.1. *"Le Louvre à Lens, un défi culturel, sociétal, économique et urbain", Profile pages, No. 110, September 2012, INSEE*

46 Able to be regenerated.

– The beginnings of the museum were largely subsidized and glorified. However, after the launch (inauguration on December 4, 2012) – which did not go without resistance – conflicts between the actors started arising; this was between the persons in culture and tourism, between appointed officials, between the locals and more distant players. There was no proper estimate of the time lapse required to produce lasting effects (truly embodied in the space). Waiting took too long, which triggered criticism or disappointment, and even anger from the actors experiencing the economic side effects (more sluggish than expected). Certain symbolic signs were noted (the opening of the museum restaurant, the Atelier, by Marc Meurin in 2013; shows offered to residents in the Museum Stage; the millionth visitor in January 2014 – that is, 14 months after the opening – the first museum visit for babies over nine months in September 2014).

– In 2015, the signing of the ALL destination contract (see section 1.7.2.3).

– The arrival at Louvre-Lens of a direction, which was both political and cultural (Marie Lavandier, who desired to implement actions rooted in the territory). Strengthening links with the Louvre-Paris, thanks to the installation of reserves in another local site. Synergies (a large meeting in October 2017), the symbolic use of the five-year anniversary after the opening and the 10 years of design, implementation and start-up, as well as a strategic projection (2019) over the following 10 years.

1.7.2.3. *ALL (Autour du Louvre-Lens)*

The destination contract was yet another means for activating the development process, as were two almost simultaneous and happy coincidences: the exaltation of places of remembrance, on the occasion of the centenary of World War I (1914–1918, therefore 2014–2018) and the promotion of mining, leading to a remarkable rehabilitation of sites which can now be "visited".

ALL (*Autour du Louvre-Lens*) was one of the 22 destinations chosen by the Ministry of Foreign Affairs to attract international clienteles. Instigated by the Departmental Council of Pas-de-Calais, ALL brought together 70 public and private partners, "committed to the tourism transformation of the mining basin", via a "contemporary" approach (in line with societal trends and novel forms of consumption). The local supply had the advantage of "being out of the ordinary", an ideal trait for differentiation. It became necessary to bring forward this coherence and territorial originality, to make tourism one of the levers of economic development. How could branding be tackled? How could actors come together to serve tourists? How could they set common goals to increase and improve the quality of the existing supply?

ALL was firstly designed to increase foreign attendance, by strengthening promotion in priority markets (Benelux, Germany, United Kingdom, as well as Japan, because the SANAA agency, designer of the museum, was Japanese, and there were Toyota premises not far away). The available means were (international and local) culture, identity values that the inhabitants were proud of, the organization of events, and creativity (to imagine the tourism of the future, its need for original experiences, its quality and mobility requirements, its use of communication technologies and virtual realities, etc.). Since March 2018, "ALL" branded products have been present in seven pop-ups in Japanese luxury department stores.

ALL also participates in a project for the promotion of jobs and competences (ADEC, *action de développement des emplois et des compétences*).

1.7.2.4. *The Louvre-Lens: an opportunity as a starting point*

The Louvre-Lens was born from a need and an opportunity. In strategic terms, there is more resemblance with the garbage can model by March[47] than with a premeditated strategy for persuading all the actors. Just as "opportunity makes a thief", the project's creators joked about their good luck, attributing it to a constellation of happy coincidences – even though some of those coincidences had been largely driven and oriented.

We could also speak of a "translation" device (in the sense of Callon and Latour). After all, contrary to what Tobelem says (to stigmatize initiative and predict its failure), any emerging project has an "above ground" side. Or rather, it is often designed by thinking about its (real or potential) roots, even though its innovative characteristics require a taming and appropriation process with a certain time lapse before achievement is reached – especially since there is an initial "distance", even an "incongruity", between the object and its context, such as Beaubourg in Paris.

In the case of Louvre-Lens (LL), the ALL destination contract was one of the elements which characterized the desire to use all the available means in an emerging context, for the sake of progress.

They quickly realized that the LL would not suffice on its own to redeploy energy in a sector hampered by complex difficulties. The title of the contract is eloquent: it is "around" (*Autour du Louvre-Lens*) that other opportunities should be spotted (while the short form "ALL" conveys the idea of bringing "everyone" on board).

---

47 This is a very common model in *management* policy. If we plan to satisfy the need for rational demonstration ex post, many decisions stem from incidental opportunities, from unexpected meetings between problem providers and solution providers.

Nevertheless, it was broadly accepted that "the LL was one of the strong levers for the territory's renewal, with a commitment from various actors to carry out joint actions and transform the territory into a tourist destination" (Grafov 2018).

While the Louvre displays achievements directly related to its territory, we should also take into account the larger events it resonates with. For example, in autumn 2020, it inaugurated a design exhibition which earned the city of Lille the nomination as the 2020 World Capital of Design.

1.8. Mélusine accueil: a system of actors with no intermediation

Mélusine accueil is a network based on the actors' geographical proximity. All members, except for one, are located inland and the majority of them, in Vendée.

	Effective members	Located in Vendée	Including those outside Vendée
Accommodation	39	39	0
Catering	10 including two off-grid	10 including two off-grid	0
Various leisure and shopping	41	36	four in Deux-Sèvres one in Charente-Maritime
Number of companies	88	83	5

Table 1.2. *The Mélusine accueil network: situation in February 2020 (source: website consulted on February 20, 2020, elaborated by the authors)*

1.8.1. *A flexible structure made up of small companies with complementary activities*

On the one hand, they are characteristic for the heterogeneity of the members who pay a membership to the association, and, on the other hand, there is a functional complementarity between them. According to the assessment we carried out on February 20, 2020 (Table 1.2), 39 members are accommodation providers, 10 members feature catering and 41 others offer activities or the sale of local products. Dissimilarities multiply within each of these generic categories. Among the hosts, there are hoteliers and owners of rooms or *gîtes*. Among the latter, members are involved in two national networks, Gîtes de France and Clévacances. Quality criteria provide a sense of unity because all members are awarded with at least three *épis* ("ears of wheat", according to the Gîtes de France rating system) or three keys. Besides, hotels are members of the Logis group. Catering options range from restaurants to *tables d'hôtes*. However, another difference lies in the fact that not all room rental companies offer this formula. Diversity is the reason for being of this association, its purpose, to provide tourists with the opportunity to program their stay as they wish, choosing from a varied supply. Alongside accommodation providers, many structures offer either leisure activities or the sale of local products. Out of the 41 various choices, only 26 offer a form of leisure activity, sometimes limited to only one season apart from summer. The other 15 proposals are essentially commercial, with boutiques at the manufacturing sites or within the region.

1.8.2. *Weak dynamics*

The association is chaired by an elected official, assisted by a six-member board of directors, but is not supported by any mediating structure. Despite the number of members tending to increase (82 in 2019), the network is losing momentum, which largely depends on the president's enthusiasm. The driving forces are running out of steam. This peculiar change may be due to the development of the Internet and the direct links that members establish via their own websites with customers. The deployment of "OTA" commercial sites (Other Transaction Authorities, or generalists like Leboncoin) calls into question such initiatives. However, the interest in being a member is not evident when reading the site. There is no provision for leverage. No tailor-made commercial proposals for customers are offered on the website. What would make sense is tariff reductions, degression, etc. for the tourists who stay at the members' accommodation in exchange for a proactive promotional approach. The heterogeneity required to match the complementarity sought by tourists in need of a global but flexible response to their needs seems excessive, particularly with regards to the role of production-oriented local crafts and purely commercial activities. These supplies offer little added value to the network, compared to those which offer more or less elaborate entertainment options (in situ exhibitions, training courses, diverse initiatory activities, etc.).

1.9. Experimenting with a tourism cluster in Morbihan

Since 2006, the General Council of Morbihan has embarked on a process for creating clusters dedicated to the department's important economic sectors: the boating and shipbuilding sector (2007), later electronics, communication, and information technologies (2008) and artistic crafts (2008). For the General Council of Morbihan, the notion of cluster is meaningful in that it complements public and private actors, thereby promoting the collective development of businesses thanks to economies of scale and resource pooling. Furthermore, the approach is also of interest for the community in terms of territorial marketing, as its helps legitimize the inclusion of the Department of Morbihan among the innovative territories.

1.9.1. *A global reflection applied to tourism*

In 2012, the president of the General Council of Morbihan, François Goulard, was concerned about the figures for the tourist economy. This department, located in the southeastern area of the French region of Brittany, benefits from tourism consumption for approximately 1 billion euros, impacting 9.5% of jobs. However, according to this actor, tourism is not evenly distributed across the territory. In fact, 25% of the department's municipalities offer commercial accommodation for tourists and are located along the coast. These 560,000 commercial tourist beds generate approximately 3 million euros in tourist tax. They are largely dominated by campsites (103,824 beds for 266 campsites). However, it is the furnished accommodation which is most present in terms of the number of establishments (1,983 establishments for only 9,228 tourist beds). Still, according to INSEE, in 2012, Morbihan had a non-market offer greater than the market offer with 79,706 second homes, 398,530 beds or 71% of the tourist beds in the department. Half of the clientele was made up of families and 80% of the tourists were French, mainly from Brittany and the Paris region. Among the foreigners, half were British. The tourist demand monitoring survey (SDT, *Suivi de la Demande Touristique*) carried out during the summer of 2011 revealed that the Department of Morbihan was in the fifth position in relation to the number of tourist overnight stays, behind the Var, Charente-Maritime, Hérault and Vendée. This finding challenged elected officials. However, the annual number of overnight stays only slightly began to change after the year 2000 (31,746,581 overnight stays in 2000, 34,179,754 in 2001, 34,745,637 in 2003 and 33,376,848 in 2011)[48]. After that, the office responsible for tourism from the General Council of Morbihan started speculating about the possibility of creating a cluster.

The General Council of Morbihan now wished to take charge of the tourist development. However, since December 11, 1978, that mission has been entrusted to the Departmental Tourism Committee, an association under the 1901 French Law, created upon the initiative of the General Council and endowed with a departmental subsidy of 80% as an operating budget. The creation of a tourism cluster was legislated upon and proclaimed during the departmental session of January 30, 2013. Its mission was the "to array all of the actors in the sector, relying on the expertise of

48 *Les chiffres clés du tourisme en Morbihan*, 2014 edition, p. 8.

the Departmental Tourism Committee on all actions relating to the territory's promotion and the engineering required for structuring this sector"[49]. Relying on its tourism department (attached to the economic development department), the General Council of Morbihan claimed its legitimacy over the Departmental Tourism Committee for managing the cluster:

> By clearly stating its tourism policy, the department is committed to its partners. The implementation of the [departmental] plan must be closely monitored and managed by the tourism service, acting as the technical representative of the departmental assembly who voted for this program[50].

By openly stating its policy on tourism, the department wished to commit to its partners. The creation of this tourism cluster matched the will of the General Council to update an orientation plan called the "Departmental Tourism Development Plan", outlined for the 2013–2015 period. It immediately showcased itself as a break with Brittany's Regional Tourism Tlan (2012–2014) by refusing to integrate a shared marketing approach[51]. We will detail how the Morbihan tourism cluster operated for two years, until the 2015 departmental elections, which put a halt to that system.

1.9.2. *The tourism cluster model for the General Council of Morbihan*

The establishment of a tourism cluster was carried out in a context of tension between the General Council and its Departmental Tourism Committee. In January 2012, an audit noted that:

> The collaboration between the General Council and the Departmental Committee was incomplete: no procedure for dialogue and regular consultation had been set up. The agreement associating the Departmental Tourism Committee with the General Council did not make it possible to define a level of articulation between the community and its essential arm for implementing the departmental tourism policy[52].

49 Excerpt from the Morbihan Departmental Tourism Plan 2013–2015, p. 79.

50 Ibid., p. 81.

51 Report of the definitive observations from Brittany's Regional Chamber of Accounts, Morbihan Departmental Tourism Committee. Fiscal years 2012 and following, March 24, 2017, p. 4.

52 Ibid., p. 19, citing the PwC audit.

As proof of these tensions, the eighth vice-president of the General Council and vice-president of the Departmental Tourism Committee was appointed as the 2013 cluster referent (instead of the fourth vice-president of the General Council, already appointed as president of the Departmental Tourism Committee). Furthermore, the change of director of the Departmental Tourism Committee in 2013 was also revealing of this animosity. Nevertheless, the General Council took control over the departmental tourism policy, justified by the emergence of a tourism cluster:

> The General Council of Morbihan has redefined its tourism development policy for the period 2013–2015 and desires to invite businesses, as well as other associative and institutional organizations related to the departmental tourist economy, to tackle a shared reflection work[53].

For its part, the Departmental Tourism Committee took note of this organization proposal and included the "active participation for implementing a tourism cluster" in its 2013 action plan. However, that initiative was not included in the financing convention with its collective supervisory authority[54].

1.9.2.1. *Preparation of the cluster via consultation workshops*

In January 2012, the Morbihan General Council appointed its communications director as head of the tourism department, who would be responsible for establishing a database of tourism actors in Morbihan. Some institutions and companies, such as the Morbihan Chamber of Commerce and Industry, quickly accessed this pooling process; whereas, others, such as the Departmental Tourism Committee, did so less promptly. With the arrival of a collaborator in the tourism department, invitations were abundantly sent to tourism companies and institutions in order to organize consultation among the tourism actors.

In September 2012, workshops were held with the goal of "proposing places of exchange so as to identify the territories' major challenges and the difficulties encountered"[55]. Coordination fell into the hands of a consulting agency, the head of the tourism department of the General Council and the director of the Departmental Tourism Committee. The participants were grouped into "professional" workshops (accommodation providers, event organizers, leisure companies) and "transversal" workshops (tourist supply and adaptation to clienteles, accessibility to offers, attractiveness of the Morbihan region, observation data, tourism and tourist reception professions). During these workshops, participants made comments and observations and envisaged different solutions. For example, during the transversal

53 Excerpt from the Morbihan Departmental Tourism Plan 2013–2015, p. 6.

54 Ibid.

55 Ibid.

workshop "tourist supply and adaptation to clienteles", it was found that the supply of leisure equipment for the "families and youth" audience was not sufficiently clear due to the large number of prescribers. A proposal was made to train seasonal workers on the supply's diversity for leisure and event sites.

These consultation workshops were allegedly implemented as a new tourism policy that reflected the departmental strategy on tourism, organized around four axes:

1) to support tourism businesses in adapting and qualifying the supply to improve its quality and perceived level of value for money, for all clienteles and all range levels;

2) to support destinations in qualifying their supply and their reception for all clienteles;

3) to guarantee an outstanding reception in order to develop the destination's attractiveness and foster guest loyalty;

4) to diversify departmental tourism to strengthen and secure its economy.

The tourism department of the General Council was designated for implementing the plan and for monitoring achievements during its enforcement. It was also to assess works as technical representative of the departmental assembly. Concerning the methodology of action, the 2013–2015 Tourism Development Plan had to define the "who does what", outline the role of the actors and "internally, define the scope of the departmental institution, leveraging the department's services and the Departmental Tourism Committee"[56]. Implementing the collaboration between private and public actors was inscribed in the tourism cluster's logic, seeking to make professionals work together and put them in contact with tourism-related sectors of intervention, as well as playing a key role in the supply proposed to the clienteles (transporters, developers, promoters, commercial unions, etc.)[57]. The consultation workshops notably brought to light the desire of tourism actors to clarify the roles between the competent structures or authorities.

1.9.2.2. *Organization of the cluster and its pilot committee*

The tourism cluster was launched on April 12, 2013 "to array all of the actors in the sector, relying on the expertise of the Departmental Tourism Committee on all actions relating to the territory's promotion and the engineering required for structuring this sector". The cluster presented itself as a tool based on the General Council's capacity to create, organize and lead a network of actors bringing together

56 Ibid., p. 77.

57 Ibid.

professionals, training organizations, financial partners, communities and tourism-related sectors of intervention. The mode of governance had to ensure the areas of competence of each actor and encourage public and private actions. The risk of engaging multiple actors could be seen as a source of competition or redundancy in the actions to be tackled and end up hindering the readability and effectiveness of departmental policies. A pilot committee for the tourism cluster was instituted in order to guarantee operationality, when implementing the Tourism Development Plan. It brought together three representative colleges:

– appointed officials with seven general advisors, including the tourism cluster referent, the president of the Departmental Tourism Committee and the arts and crafts cluster referent;

– private companies with 17 socio-professional actors (three hoteliers, two managers of outdoor hotels, one lodging manager, two managers of transport companies, two managers of tourist residences and holiday villages, two event managers, one restaurateur, one thalasso-therapy manager, one industrialist, one leisure site manager, one travel agency manager);

– partners with 11 representatives of consular, institutional or federative structures (Chamber of Commerce and Industry, Tourist Offices, Regional Council, Chamber of Trades and Crafts, etc.).

Therefore, the members of the board of directors of the Departmental Tourism Committee were not de facto members of the pilot committee. Some persons belonged to it (on a personal basis), but the engagement of "network leaders" representative of branches and professions was not systematic, unlike what happened with the board of directors of the Departmental Tourism Committee.

On this topic, one of the cluster leaders declared:

> We were looking for the most suitable persons capable of having ideas, even if these were a bit far-fetched, the goal was to differentiate ourselves from the classic board of directors and to stimulate new ideas. The idea was to focus on people who had a slightly different vision, either by company size or the personality of their managers[58].

The pilot committee, led by the General Council, validated the actions proposed by the consultation workshops. The cluster referent, a departmental advisor, defined the cluster's purpose as a desire to:

58 Interview conducted at the tourism department of the General Council of Morbihan, in the city of Vannes on November 4, 2013.

> get to know the expectations of professionals, for everyone to participate. The idea is to strengthen tourism culture within the General Council by being innovative and competitive in market segments, on certain themes. As for the articulation with the General Council, everyone is expected to get involved. Four orientations are given for the renewal of the Tourism Development Plan which need to be taken up by the cluster. The Departmental Tourism Committee will be in charge of implementing an operational work, particularly in terms of marketing[59].

The pilot committee was responsible for monitoring the reflection encouraged by the thematic workshops in compliance with the four orientations set by the 2013–2015 Tourism Development Plan. Meetings were held three times a year in assembly. During these sessions, the spokespersons for the thematic workshops presented the actions depending on their progress and the pilot committee issued an opinion. For example, during the pilot committee meeting on October 4, 2013, the president of an accommodation network in the department presented the various courses of action discussed and finally implemented within the thematic workshop "family and proximity" (which falls under orientation #3); and, in particular, a *benchmarking* action on the sector targeting specific clienteles. The following step would involve developing a quality charter with an audit of volunteer professionals, and then elaborating a guide. During the meeting, a tourist office director, member of the pilot committee, intervened to clarify that the "family label" already existed and that there was no point in reinventing something that was already properly working. The assembly re-discussed the term, specifying that it was not really a "label", but a charter of commitment applied to establishments, not to municipalities or tourist offices. At the end of each pilot committee meeting, the actions to pursue were decided upon, together with the actors to be engaged. Therefore, at the end of the pilot committee of October 4, 2013, one thematic workshop ("business tourism"), initially piloted by the tourism department of the General Council, was taken up under the management of the Departmental Tourism Committee in connection with the Chamber of Commerce and Industry. Also, two thematic workshops continued to be managed by the tourism department of the General Council ("family and local customers" and "tourist signage") and five new workshops were scheduled, still under the management of the General Council's tourism service ("overall satisfaction", "good reception practices", "tourist promotion of events", "tourist consumption of secondary residents" and "reception of groups and coach operators").

---

59 Introductory speech by the vice-president of the General Council of Morbihan, referent of the tourism cluster during the opening of the pilot committee on October 4, 2013 in Vannes.

1.9.2.3. *Coordination of thematic workshops*

After the launch of the tourism cluster in April 2013, thematic workshops were organized, led by the tourism department of the General Council, in which any tourism professional from the Department of Morbihan could take part. In 2013, five workshops[60] took place 15 times at different places (in the General Council premises and in tourist establishments). On average, each workshop session gathered 10 persons from diverse backgrounds and distinct professions, in order to discuss the themes defined in accordance with the main orientations of the Tourism Development Plan; and, in particular, the ongoing projects. The choice of workshop themes was the result from the consultation workshops, having made it possible to define the 2013–2015 Tourism Development Plan. As one of the tourism facilitators from the General Council explained, "the idea is for professionals to act as requesters, for us to count on their strength and their mobilization to change things, because we [the facilitators of the General Council's tourism office] cannot do so"[61]. Each workshop session was preceded by a reminder of the cluster's logic as a place for sharing, inviting the participants to give testimony of their experience, rather than having information passed down by an institution. The goal was to organize joint actions between public and private actors, as well as among the private actors themselves. The facilitators reminded that "putting oneself in cluster mode" requires adaptation:

> When I launched the workshops, I insisted on the fact that when we operate in a cluster, it is clear that we are competitors (two hoteliers for example), but even so, this is a place of reflection where we put the collective at the service of everyone. This is what we should cultivate, making it clear that the collective is superior to the individual and once this is properly understood, the participants will realize that by serving the collective, they are also doing a service to themselves, at the individual level. The idea is to reverse the powers, to embrace sharing, solidarity, alliance strategies, all in tune with the new times[62].

The positions adopted by the members of the Departmental Tourism Committee during these workshops could make relations complex. However, certain participants

---

60 Five workshops were organized: "targeting specific clienteles", "outstanding hospitality", "qualifying tourist offers and networking with tourism actors", "business tourism", "secondary residents".

61 Interview conducted at the tourism department of the General Council of Morbihan, in the city of Vannes on November 4, 2013.

62 Interview conducted at the tourism department of the General Council of Morbihan, in the city of Vannes on November 4, 2013.

who were traditionally considered as "opinion leaders" preferred to let other actors express themselves.

The tourism cluster experiment led by the tourism department of the General Council of Morbihan ended in 2015, the year of departmental elections and the promulgation of the NOTRe law. First, elections changed the presidency of the Departmental Tourism Committee. On March 29, 2015, the departmental elections were over. The General Council of Pont-Scorff, president of the Departmental Tourism Committee since 2004, did not stand for election; so, in 2015, a new president arrived as the head of the departmental tourism institution. Therefore, the former vice-president of the Departmental Tourism Committee and former referent of the tourism cluster took full presidency of the Departmental Tourism Committee. Afterwards, the law of August 7, 2015 (which established a new territorial organization for the Republic[63]), came to modify the competence of communities, and in particular the department's economic competence, since it removed the general competence clause and the capacity of the departments to intervene in the economic development area. The department would maintain its orientation as a guarantor of human solidarity and territorial cohesion. Also, on November 1, 2015, a new territorial development department was created at the General Council of Morbihan, which merged the old economic and territorial development departments, and removed the tourism service and its tourism cluster, which would therefore remain in the state of experimentation for the General Council of Morbihan.

1.9.3. *Morbihan's tourism system for the actors*

The experimentation of a tourism cluster in Morbihan made it possible to identify the functioning of tourism actors, in terms of their professional relationships and their bonds with others. Therefore, the Morbihan tourism system was thoroughly revealed during the thematic workshops, a place for exchanges, testimonies and problem identification.

1.9.3.1. *Stakeholder mobilization and institutional tensions*

The thematic workshops organized during 2013 mobilized a wide variety of professionals. There, we could find directors and presidents of tourist offices from seaside, urban or rural destinations, hoteliers from independent establishments, hotel chains of heterogeneous ranges, castle owners, managers of museums or leisure sites, etc. During a workshop on the theme "quality of the tourist supply and networking" organized at the General Council in Vannes on June 6, 2013, the

63 Law No. 2015-991 of August 7, 2015 pertaining to the new territorial organization of the Republic, published in JORF No. 0182 of August 8, 2015, p. 13, 705.

participants stated their motivations for participating in this cluster as the session started:

> In inland Brittany, we need an external perspective, to improve quality, we are looking for a critical perspective on how to draw visitors in (Castle owner).
>
> How can we ensure complementarity between the tourist office, the department and the Departmental Tourism Committee? Redefining goals in terms of complementarity instead of competition, when we listen to certain directors [of tourist offices], well, it is quite shocking (A tourist office director).
>
> Working in synergy, complementary work, finding a leader to foster networking (A tourist office director).
>
> Identifying the role of each (public and private) actor, creating synergy between the actors of tourism in Morbihan (A communications manager of a leisure site).
>
> I am glad that there is a cluster, it lets us outpour our worries, find instructions, solutions (A director of a travel agency).
>
> Better organization, work on quality, production by workshops and information coming back down, we don't all have a similar vision of what constitutes quality (A tourist office director).
>
> Carefully chosen activities, good events that work well, we also have expectations about New Information and Communication Technologies (NTICs). We are also concerned about defining "non-quality" because we know quality well (A manager of a leisure facility).
>
> In the search for overall information, networking (A museum manager).
>
> There's no point in being alone, networking means that tourists are freed from any barriers (A campsite owner).

Beyond the search for support through the network, the actors' problems differed, depending on the size of their structures and their experiences. On the contrary, the actors identified the existing tensions between the General Council and the Departmental Tourism Committee, which led the facilitators to make the following observation:

> We clearly feel that the scope, and the articulation are not yet completely fluid. In the coming weeks we will have a clearer panorama, bearing in mind that we are positioning ourselves as an incubator and a business nursery, and are not tied to a marketing or promotion strategy[64].

Moreover, the actors felt this difficulty. As evidenced by an accommodation director: "the establishment of the cluster by the General Council was a stone in the shoe of the Departmental Tourism Committee"[65]. A tourist office director declared:

> When the department's tourism service was formed, we had a little difficulty understanding the difference between the role of the Departmental Tourism Committee and the General Council. At first, we imagined that the department desired to regain control of the Departmental Tourism Committee (DTC) and restore coherence, but then we saw that there was another vice-president of the General Council, referent of the cluster and a new director of the DTC who did not want to hear about the cluster. there was certain confusion in terms of information[66].

The lack of coordination between the services of the General Council was also pointed out by a tourist office director: "politically, within the General Council's 'boutique', it was accepted that the president of the Departmental Tourism Committee did not agree with the General Council's policy on the cluster, it is absurd"[67].

1.9.3.2. *Local and regional policies versus departmental policy*

The Department of Morbihan is a territory where the concentration of facilities and tourist numbers is located along the coastal fringe, revealing an imbalance (from the point of view of the General Council). However, during the workshops and pilot committee meetings for the tourism cluster, it became clear that the department was also being held responsible for a lack of will to publicize tourism. In that sense, a tourism professional from Lorient denounced the historic rivalry between Vannes and Lorient:

> From Lorient's viewpoint, it is difficult for us to work together with Vannes. It is a historic fact, even in terms of supply, Lorient has often

64 Interview conducted at the tourism department of the General Council of Morbihan, in the city of Vannes on November 4, 2013.

65 Interview conducted on March 3, 2015 with a tourist accommodation provider.

66 Interview conducted on March 9, 2015 with a tourist office director.

67 Ibid.

> been considered as a somewhat secondary territory rather associated with fishing and the naval […], today we have the impression that the General Council is much more interested in the development of the region of Vannes than in Lorient. There are political choices, I can understand that, but a department is a department[68].

From this perspective, despite its management by the community in question, the cluster had the merit of encouraging meetings between the actors and "of gaining further knowledge about what my neighbor is doing, having a complementary supply to what I offer. The goal is for the economy to develop at a territorial scale", declared the director of a tourist office[69]. The question of political power also came into play. In the case of Lorient, it turns out that the municipality and the agglomeration had a "left-wing" political sensibility; whereas, the General Council had a "right-wing" orientation. In broader terms, there was also the question of territorial power. Each appointed official sought to stand out, for which tourism constitutes a strategic means. For example, in the Brocéliande forest, all the communities of municipalities had monopolized the tourism competence (some of them for over 30 years); whereas, this was optional until 2015, to ensure tourism development in their territories. Another example: when the Brittany Regional Tourism Committee created the "Brocéliande destination" straddling the departments of Morbihan and Ille-et-Vilaine – the first tourist destination inaugurated by public and private actors on March 18, 2013 – the Morbihan General Council did not adhere to the tourist destination strategy of the Brittany Regional Tourism Committee. It should be noted that the Brittany Regional Tourism Committee classified the Breton departments according to their tourist attractions. In particular, Morbihan was divided into four destinations: "Bretagne Loire Ocean", which brought together the municipalities of Ille-et-Vilaine, Loire-Atlantique and some of Morbihan (La Roche-Bernard, Penestin), "Bretagne Sud Gulfe du Morbihan", bringing together Lorient and Vannes, "the heart of Brittany kalon Breizh", reuniting the municipalities from the departments of Finistère, Côtes-d'Armor and Morbihan, and finally, "Brocéliande". The actors somehow felt a confrontation between several policies, as expressed by the director of a tourist office: "it all comes down to a question of local power, local structures want to have power, be them the elected officials or technical actors. In tourism, power plays a prominent role"[70]. The interest of the cluster was perceived as a means for creating a participatory intermediate level between local power and departmental power: "animating a network like the cluster is anything but being in power", explained a director of a tourist institution, "in any case, we are faced with an ego problem", explained a campsite manager[71].

68 Interview conducted on March 25, 2015 with a leisure site equipment manager.

69 Interview conducted on March 9, 2015 with a tourist office director.

70 Interview conducted on March 9, 2015 with a tourist office director.

71 Interview conducted on March 3, 2015 with a campsite manager.

1.9.3.3. *Networks of belonging to break free from territorial and organizational impositions*

In the face of a heterogeneity of public and parapublic counterparts, who carried out their actions on multiple territorial levels and acted on behalf of their territories in tourism development and promotion, the tourism professionals in Morbihan had to become organized. Proximities were established by sector, by territory, by theme, which facilitated the creation of networks. Within these networks, professionals exchanged information and knowledge they considered necessary, because they voluntarily adhered to these networks. As a tourist destination, "Absolument Brocéliande" brought together approximately 50 professionals, without a legal structure, "it is just a network which gathers professionals. Its intention is not to create immediate economic benefits, we operate on exchange, on the sharing of values. We are more in tune with development than with promotion", declared one of the members[72]. On the coast, seven tourist offices in the Auray region also created an association in May 2010. The goal of the "Group for the promotion of Quiberon Bay and its islands" was to publicize their destination. At the time, the directors of the tourist offices sought to go beyond municipal limitations, foreshadowing collective competence by the Auray agglomeration community. Other tourist offices on the Morbihan coast (those in prominent tourist destinations) shared their reflections with 20 other tourist offices in seaside resorts in Brittany via the "Sensation Bretagne" network. For them, "the idea was to globally promote the coastline, including a coordination component and a digital strategy, given the fact that the big seaside resorts had a locomotive effect on the small destinations", stated the director of a tourist office[73]. The actors also grouped up by sector. This was the case of the Union of Hotel Trades and Industries, the National Federation and the Breton Union of Outdoor Hotels. Some networks were even supported by the Departmental Tourism Committee. This is what happened with the Departmental Association of Leisure Facilities, who were subject to a partnership agreement in 2006. For leisure equipment professionals, "it was an important network, as it provided key knowledge and launched work sessions around common themes (depending on the topic chosen). One may attend the meetings, or not", declared the director of a leisure facility[74]. Finally, another form of gathering (the thematic network) mobilized actors around a common subject or a shared value. This was the case of sustainable development, which brought together 50 professionals from Morbihan, transforming the former professional tourist group from Auray and the Quiberon Bay region called "Morbihan Sud" and created in 2002, into the "Morbihan Responsible Tourism" association. In 2015, the president justified the interest of the association for tourism professionals:

72 Interview conducted on March 9, 2015 with one of the members of the "Absolument Brocéliande" group.

73 Interview conducted on Tuesday, March 3, 2015 with a tourist office director.

74 Interview conducted on March 25, 2015 with a leisure site equipment manager.

> The name "Morbihan Responsible Tourism" is more in line with the actors' approach. It is a professional network, there is an adventure circuit which is very well anchored in this approach, there is also unusual accommodation, or a more classic campsite. The old name "Morbihan Sud" focused too much on the coast. In addition, it did not mark our environmentally friendly footprint, and establishments in the north of the department did not want to join a network called Morbihan Sud[75].

This network was also part of broader networks. Therefore, the president told us that the eight members of the association's executive committee were involved in local structures and local meetings (development committees of the region of Vannes, of the region of Auray, the communities of municipalities and agglomeration, the TGV club) with the aim of publicizing actions and keeping their audience informed about innovations, in order to stimulate and trigger new projects. As such, a "car-free vacation" project was launched with the actors, in order to test other modes of travel.

The thematic workshops of the tourism cluster also promoted a rapprochement between professionals in the business travel sector. At the end of the thematic workshops reuniting event and seminar professionals, the "Morbihan business" association was born in 2014 under the impetus of the General Council's tourism department, and then thanks to the support of the Departmental Tourism Committee and the Chamber of Commerce and Industry. In 2019, during the change of presidency, the network included a total of 49 members, who hoped that the association would position the Morbihan destination on the "business" market, promote the sector, engage in the qualitative improvement of the supply and facilitate exchanges and synergies. It should also be noted that from 2016, the Departmental Tourism Committee and the Morbihan Responsible Tourism association prepared the creation of a brand and an association called "Green Morbihan", which materialized after four working meetings in March 2017. Morbihan Responsible Tourism embodied a cluster dynamic itself, since the association claimed to be a nursery and support for professionals engaged in a sustainable development approach. The association played an incubator role for the Green Morbihan association[76]. The terms "business nursery" and "incubator" clearly indicate a cluster approach.

---

75 Interview conducted on March 3, 2015 with the president of the Morbihan Responsible Tourism Network.

76 Anonymous (2017), "Green Morbihan, la démarche s'organise!" [Online]. Available at: https://www.morbihan-pro.com/green-morbihan-la-demarche-sorganise/ (accessed: March 9, 2020).

1.9.4. *Lessons from a local authority's attempt at "clustering" tourism*

The initiative from the General Council of Morbihan to implement a cluster approach in 2012 seemed original, as no other French department had previously embarked on such an adventure. The principle was coherent with a cluster's DNA, something amply known in the industry. However, its inner workings, governance and spatial dimension quickly showed the limitations of having a local authority manage the project at a departmental level, while bringing to light the expectations of tourism professionals regarding networking relevance.

1.9.4.1. *Involvement and time granted to the collective*

While the tourism cluster experiment lasted, we could note that very heterogeneous actors took part in workshops and the pilot committee meetings. These elected officials, technicians, managers, owners from various professional and geographical backgrounds saved part of their time for collective reflection. As one tourist office director declared, this openness was also sought by the participants:

> What I find interesting in the cluster is that it is open to everyone, without necessarily representing any interest in particular, one man, one voice, it's like a forum, we share ideas, it's a real asset. If we don't reach the majority of the opinions, we drop the subject[77].

However, the diversity of profiles during the thematic workshops also presented the difficulty of giving voice to actors who were more or less trained in the tourism sector, with a deeper or poorer knowledge in the trade. The director of a leisure facility told us:

> Between the intention, the idea and its accomplishment, there is a relatively high step. We are confronted with a diversity of entities (in legal structure, in size, in the sector of activity), we are dealing with issues and structures which do not share the same approach[78].

During the thematic workshops or pilot committee assemblies, it became evident that the actors who mastered "the codes and principles" of the institutional or industrial tourism sector mainly strove to intervene (even almost inopportunely), on the grounds of a claimed expertise. However, during the presentation, the discourse manifested a desire for participants to express themselves at an individual level:

> At each workshop I projected a slide explaining that this is a space for respect, for cooperation, kindness. But there were times when it all got

77 Interview conducted on March 9, 2015 with a tourist office director.

78 Interview conducted on March 25, 2015 with a leisure site equipment manager.

> out of hand. I warned more than one reminding them that we had to be careful and tune ourselves in a benevolent mode. The 'all we have to do, it suffices to…', merely showed that some persons are used to being know-it-alls, that they find it difficult to see what others are doing or to call themselves into question[79].

This situation took place during a "business tourism" thematic workshop in September 2013, when an institutional actor, with an academic training in tourism and experience in multiple positions, wanted to take the lead and impose his actions on the sector. Several years later, a facilitator still remembered the situation:

> At the meetings I attended, this actor in question always operated like this: "this is my suggestion, either you take it or leave it!" The cluster's logic does not work that way. We prefer to give the participants a topic and for them to work on it. This actor in question had created an unspeakable mess during this workshop and the people who took part in it were disappointed in not being able to successfully carry out their actions, they were unable to work correctly as they had during the previous workshops[80].

This situation revealed the participants' interest in getting to know each other better and getting out of an "individualistic" game. As one accommodation director explained: "the problem is that we sell the same thing, in the same territory, and we can't organize ourselves to prevent duplicates and know how to invest where needed"[81]. Therefore, the cluster became useful for actions in which the professionals felt lacking, or deprived from the means for seeing their projects through. Moreover, volunteering was required in this type of approach, as a member from the *Absolument Brocéliande* group told us:

> What I find remarkable is that we have professionals who are willing to spend time on exchanging information, through educators, work commissions. Actually, they devote their time without expecting immediate economic benefits[82].

However, we also noted that many of the actors who participated in the cluster's thematic workshops were actors who already devoted time to various networks or

79 Interview conducted at the tourism department of the General Council of Morbihan, in the city of Vannes on November 4, 2013.

80 Ibid.

81 Interview conducted with an accommodation director on March 27, 2015.

82 Interview conducted on March 9, 2015 with one of the members of the *Absolument Brocéliande* group.

groups. An accommodation provider with networking experience considered that the time dedicated to the collective was beneficial for them and helped them in finding new ideas:

> The tourism cluster is the voice given to professionals. This is something that hasn't been officially done before. I have always railed against professionals who say that we do nothing for the other professionals. I believe that I need to include these exchange times in my schedule. During the low season, it only takes 3-4 days per month for attending all the networks. In terms of time, it is not important, what is important is the content[83].

For others, choices must be made among multiple requests, as evidenced by a tourist office director:

> There are only three of us at the tourist office, so in terms of implications we have to make choices and it is not always easy. I myself belong to the technical committee for the destination with five other tourist offices. I have to spare one day per month, when two meetings are held[84].

The motivation to participate in these meetings is also that actions are really fruitful. For a tourist office director who did not attend any tourism cluster meetings, the problem was "meetingitis", the attitude of "holding meetings, for the meetings' sake":

> I find that the more we communicate, the less effective it becomes. I sincerely believe we cannot hold any more meetings. Sometimes we attend them because it's politically correct to do so. But I have a lot of work, so if I spend half a day and don't get anything out, it's a lost day. I hope that we can come up with something concrete. These meetings are super time-consuming, and when you read the minutes, you have the impression that we are stating truths, that we are knocking down open doors[85].

A director of leisure facilities suggested coordinating the public actor and the private actor schedule, particularly in terms of deadlines and project achievement:

> We need to identify needs, have a roadmap to identify common needs and set goals and deadlines that are not too ambitious. The cluster will

83 Interview conducted with an accommodation director on March 27, 2015.

84 Interview conducted on Tuesday, March 3, 2015 with a tourist office director.

85 Interview conducted on March 25, 2015 with a tourist office director.

> prove its usefulness and its actions will become trustworthy. For transformations to occur, the cluster's actions must be successful[86].

One solution would be to encourage spontaneous groupings to discuss current topics, as recommended by a leisure facility director:

> If we want to be effective, we have to work more finely in groups (regardless of whether we are less generalist with equivalent structures around a common topic, or if we have a similar size or common operating issues). We could find rich exchanges in communication, exploiting the other's point of view[87].

1.9.4.2. *Legitimacy and skills*

Power games were established between the actors of tourism. However, not all actors were competent for the same actions. The cluster principle intended to bring together multiple actors. However, the question of the legitimacy for carrying out actions and piloting the cluster disrupted the proper functioning of this experiment.

The conflict between the tourism department of the General Council and the Departmental Tourism Committee around the cluster tool revealed the perception that the actors had a preeminent role as historic tourism institutions. For many tourist offices, it was the Departmental Tourism Committee who should launch that system: a tourist office director told us:

> I think this is the role of the Departmental Tourism Committee, which is a tool of the General Council at the service of tourism. This took me by surprise at the beginning, when we all realized that there was greater engagement on the part of the General Council regarding the cluster initiative.

Another tourist office director even called into question the competence of the General Council to carry out such a system when the Departmental Tourism Committee was already legitimate:

> I am a little doubtful about all these initiatives which always begin with a very positive feeling, but ultimately run out of steam, or are not followed by any concrete action. Perhaps a more recognized leader is missing. When I looked at the reports, I said to myself, well, that is the job of the Departmental Tourism Committee, why are we still

86 Interview conducted on March 25, 2015 with a leisure site equipment manager.

87 Ibid.

> coming… I think that the Departmental Tourism Committee is sufficiently mature in Morbihan to ask these questions[88].

However, this same director acknowledged having participated in the creation of a network of tourist offices in the region to compensate for a failure in the promotion and communication mission of the previous management by the Departmental Tourism Committee:

> It's true that at the time we had a small problem with the Departmental Tourism Committee, because we were not fully satisfied in terms of promotion. It is true that there was a time when there were practically no trade shows, promotion did not seem to be at the heart of the activity. A need was thus expressed. Now we do a lot of operations with the Departmental Tourism Committee, that is to say, the Departmental Tourism Committee has changed direction[89].

For others, the General Council such as the Departmental Tourism Committee sought to assert their legitimacy in the face of a specter of territorial reform, which called into question relevance at a departmental level. A tourist office director interviewed in March 2015 seemed pessimistic about the cluster's future, particularly if the Departmental Tourism Committee were to take over:

> Does the cluster have a future regarding the open conflict with the Departmental Tourism Committee and the General Council? I'm not sure… Is the Departmental Tourism Committee committed to the cluster's culture? In my opinion, running a network like that is anything but being in power and the Departmental Tourism Committee is currently in power[90].

For a transport company manager, the problem rather lied in the General Council's motivations to engage the actors:

> The General Council has a political vision of things. We are on the field, we see what is happening. So, the cluster allows us to meet, that's good, but we have to stop working for the General Council, which is going to come up with yet another policy with key funding ideas subject to their good will[91].

88 Interview conducted on March 25, 2015 with a tourist office director.

89 Ibid.

90 Interview conducted on Tuesday, March 3, 2015 with a tourist office director.

91 Interview conducted on March 26, 2015 with a transport company manager.

As the director of a leisure facility explained to us, the duplication of actions also made actors lose any hint of legitimacy:

> Structures like ours, we have our own *marketing* engine. What is the complementarity with a tourist office or a Departmental Tourism Committee? Well, it turns out that we know them well and we work on good terms, but in all of this, isn't there a loss of means in the goal to be achieved?[92]

The difficulty of experimenting with the tourism cluster also lay in providing knowledge to less professional actors, who were looking for other actors in the same situation through this approach. As a tourist office director emphasized, the provision of expertise seemed necessary:

> We cannot do without support at the local or the cluster level. And when you are in this type of operation, it is very important to have expertise. Things are going extremely fast in the tourism sector, and we have a lot of professionals who are not aware of reality[93].

Finally, between information, exchanges, decision-making and actions, the articulation of actors required consensus and, once again, volunteerism. In November 2013, one of the leaders of the tourism cluster explained to us the difficulty on this point:

> The relationship with the Departmental Tourism Committee is currently complicated, because it is not easy to understand that we were commissioned by the General Council to work on strategic reflection, on the engagement of public and private actors. We have positioned ourselves as an incubator and a nursery. Then, the promotion mission falls under the scope of the Departmental Tourism Committee, which continues the production chain.

However, following the creation of the departmental tourism agency in February 2019, the agency offered services to businesses or to the territory: the creation of tourism projects; tourism development in the territories, positioning the product/project; valorization of the offer and product marketing. The new Morbihan Departmental Tourism Committee thus embodied far more than a promotional mission as had been imagined by the tourism department of the General Council.

92 Interview conducted on March 25, 2015 with a leisure site equipment manager.

93 Interview conducted on Tuesday, March 3, 2015 with a tourist office director.

In conclusion, the “tourism cluster” approach tested by the Morbihan General Council promoted opportunities for discussion among tourism professionals. The Departmental Tourism Committee, which benefited from a mission by this same General Council, considered itself deprived of its activities, with a feeling of dispossession regarding the networking of tourism actors. The cluster’s trajectory was ultimately marked by the workshop participants, insofar as they did not feel targeted, or had expectations surpassing the cluster’s own skills. The 2015 departmental elections were a pretext to put an end to the cluster system, which had never really worked following a top-down approach. Tourism professionals did not need public actors to act. Conversely, these professionals were structured around a multitude of networks. The bottom-up approach would have been more favorable for the cluster’s formation.

1.10. Saumur and its region: making the territory an internationally recognized tourist destination

Straddling the Loire in the Department of Maine-et-Loire, since the end of the 1980s, Saumur has innovated the tourism competence organization over a large area around the town. The destination is set within a larger tourist area, the Loire Valley. The boundaries of this space are difficult to determine, because the river crosses a set of alluvial plains from Roanne to Nantes, constituting a natural unit. Until the 19th century, the Loire River was used for the transport of wood, hemp, wine. During the 19th century itself, it encouraged artistic diffusion, introducing “Renaissance architecture, art, gardens, as well as luxurious industries such as silks and earthenware” (Dion 1978, p. 10). At the beginning of the 20th century, navigation on the Loire disappeared, giving way to tourist frequentation of the main towns, castles, inherited gardens, enhanced by tourism (Boyer 2005). In 2000, the term “Val de Loire” was included on UNESCO’s Natural and Cultural World Heritage List. The area extended from upstream Sully-sur-Loire to downstream Chalonnes-sur-Loire (Petit and Sanguin 2003). Loire tourism is established outside major concentrated developments, and promotes numerous small towns (Bonneau 1975), thus ensuring a polarizing effect. This is the case of Saumur, which, associated with a few towns and villages, constitutes the Saumurois.

1.10.1. *Saumur: a town located along cultural and tourist routes*

Since the 17th century, the town of Saumur has been located over the Grand Tour routes. Along the routes followed by these young British people towards Italy, halts were made in France, in particular, in Blois, Tours, Saumur, necessary pauses along a long route (Pittion 2014). Then, in 1843, the first trains began to run between Paris and Orléans, between Orléans and Tours in 1851, and Tours and

Nantes in 1861. This phenomenon encouraged the circulation of tourists towards a destination characterized as "the Loire castles". Since the beginning of the 20th century, photographs and promotional posters of railway companies have contributed to the promotion of this destination (Bertho-Lavenir 2007). Among the sites referenced for promoting the Loire castles, the Orléans railway company included Saumur as part of the automobile circuits offered in Touraine and the visits to the castles on the banks of the Loire. In the 1950s, the first wild campers were recorded in the meadows of the Loire at Dompierre-sur-Loire and Panay (Bonneau 1975). In the 1970s, it was the cellars and troglodytes that prompted tourism development (Rewerski 1983, p. 568). Certain small towns close to Saumur also showed a certain dynamism in tourism enhancement. Some examples are Montreuil-Bellay or Montsoreau, which created an initiative union in 1968 (Bonneau 1973, 1975; Violier 1999). Saumur has a remarkable tourist location. In a survey carried out in October 1998, Laurence Moisy was the first to stress the fact that Saumur brings together sites recording the highest numbers of visitors in the Department of Maine-et-Loire: "the Fontevraud Abbey and the Doué-la-Fontaine animal park exceed 160,000 visits. The troglodyte village of Rochemenier, the museums of Saumur, the large cellars welcome several tens of thousands of visitors" (Moisy 2001, p. 110). Then, in comparison with Laval and Le Mans, she observed that "the share of visitors on tour in Saumur is far higher than the visitors in transit, which is not the case for the other two cities" (ibid., p. 109). All things considered, traveling to the Loire castles can be organized as an excursion from Paris or as a series of strategic halts along the Loire (Lozato-Giotart 1987; Mirloup 1987). Besides, the Saumurois is set within a "Loire Valley" itinerary, where Saumur constitutes a link town, a multipolar space with a cumulative dynamic (Moisy 2001, p. 109). In this sense, we may better appraise the call for a structuring of the "Saumurois" space by Nantes geographers in 1968: "Saumur must be the capital of a complex extending from Candes to Saint-Mathurin, radiating across the Saumur plateau and across the Baugeois" (quoted by Bonneau (1975, p. 26)).

1.10.2. *The* Saumurois*: a hub inscribed within a regional space*

In the 1970s, structuring a tourist area around Saumur was not so obvious, as evidenced by the failure to create a structuring hub, bringing together the towns on the banks of the Loire to the west of the town. Despite the creation of an association "Carrefour Anjou-Poitou-Touraine" in 1975, the east of the town remained largely "unorganized" (Bonneau 1975, p. 26). However, indicators on tourist numbers revealed the interest in structuring the Saumur region. The duration of tourist halts in Saumur reached up to two days, longer than usual for a town far from the coast. Convinced of a favorable environment and notable attendance, the town's actors and the director of the tourist office developed a tourist basin strategy (Violier 1998b, p. 334). In 1989, with the help of the Region, an international tourist hub was created within the framework of a state-region plan contract, in view of promoting a coherent territory to match tourist practices. The Saumur International Tourist Hub (ITH), with an associative status, brought together 18 members, including the Saumur tourist office, the Chamber of Commerce and Industry, the two rural regions of South Saumur and the Authion Valley. This tourist basin's territorial base fell within strict institutional boundaries (Violier 1998b). The adopted strategy envisaged a connection between urban actors, relying on rural initiatives in order to create a tourist area in which tourism would have a preponderant role. In 2015, the annual budget of the Saumur international tourist hub amounted to €400,000, including €198,400 from its four public intermunicipal cooperation establishments (EPCI): the Saumur agglomeration (79.6%, i.e. €2.157 per inhabitant), Doué-en-Anjou (6.7%, corresponding to €0.886 per inhabitant), Gennes (4.1%, i.e. €0.886 per inhabitant), and Longué (9.6%, i.e. €0.886 per inhabitant). The goal of this association was to promote a territory more adapted to tourist practices, ranging from the gates of Angers to Touraine and extending beyond the boundaries of the town of Saumur.

Figure 1.10. *Schematic map of the territory of the Saumur-Val de Loire Tourist Office (Piriou 2019, p. 133)*

1.10.3. *Reorganization of Saumur tourism institutions through public policies*

The original approach implemented in Saumur made it possible to organize cooperation between different actors, facilitating a new territorial division. The French law of August 7, 2015, relating to the new territorial organization of the

Republic (known as the NOTRe law), attributed the tourism competence to the agglomeration. Therefore, the tourist office was summoned to cover the perimeter of the intermunicipal cooperation public establishment (EPCI). The Saumurois had anticipated these changes in perimeter, since in 2001, a community tourist office was created. This led to the dissolution of the associations which corresponded to the tourist offices of neighboring municipalities (Brain-sur-Allonnes, Fontevraud, Montreuil-Bellay and Le Puy-Notre-Dame), to the benefit of the Saumur tourist office, which endowed them with the status of branches. In that context, the international tourist hub disappeared, as the inter-municipalities had acquired full competences in tourism. In the two years following the NOTRe law, several inter-municipalities shared the development and tourism promotion of Saumur: the regions of Doué-la-Fontaine, Gennois, Loire-Longué and Saumur Loire Développement. Certain inter-municipalities also benefited from sites with the best attendance figures in the Department of Maine-et-Loire (218,523 visitors to the Bioparc zoo in Doué-la-Fontaine, 208,820 visitors to the Royal Abbey of Fontevraud). However, Saumur Castle only had 75,000 visitors. Note that in 2014, the site was only open between April and November.

On January 1, 2017, the perimeter of the inter-communality that integrated the town of Saumur was extended, including the former communities of municipalities of the region of Doué-la-Fontaine, Gennois and Loire-Longué. They all became grouped under the "Agglomeration Community of Saumur-Val de Loire", with a total of 49 municipalities. The tourist offices of Doué, Longué and Gennes (which had already been working since the end of the 1980s with the tourist office of Saumur, within the framework of the Saumur International Tourist Hub), all merged to create the community tourist office of Grand Saumur (Figure 1.10).

1.10.4. *Defining the role of a new tourist office*

The attribution of the tourism competence to the EPCI modified the historical missions of the association responsible for the tourist office. Until January 2017, the region's tourist offices were responsible for reception, information, promotion and marketing via the international tourist hub. Some of them were also in charge of organizing outdoor activities and to support hiking. The new tourist office was given new missions relating to engineering and development, the networking of actors and sectors, the coordination of events and activities, and the support of tourist providers, as well as monitoring. The exchanges carried out within the framework of the Saumur International Tourist Hub, innovative for the time, made it possible to create collaborative reflexes between the actors. However, the public authority moved towards the status of a local public company, created by deliberation of the municipal council on March 31, 2017, then of the community council of the Agglomeration Community of Saumur-Val de Loire on April 6, 2017. Among the actions tackled by the intercommunal tourist office, was the competence to organize and manage an annual event: the Anjou vintage Vélo. Bear in mind that to achieve its goals, the tourist office increased its permanent employees from 19 to 25, including 23 full-time equivalents.

1.10.5. *Governance of the public company: collaboration between actors to the detriment of the territory used by tourists*

Under the public company's status, only the local authorities (which were also shareholders) had the total control over governance, managing a budget of approximately 1.4 million euros. Elected officials shared seats on the board of directors according to the distribution of capital between the communities (the Agglomeration Community of Saumur-Val de Loire, majority shareholder at 72.22%, the town of Saumur, shareholder at 16.66% and the municipality of Doué-en-Anjou, shareholder with 11.11%). However, the management bodies expressed the wish to create a technical committee comprising 40 socio-professional members (wine union, hotel club, winegrowers, accommodation providers, etc.).

Nevertheless, with regards to tourist practices, the perimeter of the Agglomeration Community of Saumur-Val de Loire (defined in 2017), was not the most suitable. Tourists identified the Loire castle destination, and the names of the old provinces "Anjou", "Touraine", with the territorial denomination around the town of Saumur (Piriou 2012). Despite this, since the end of the 1980s, political actors (region, then inter-communities) had sought to differentiate the Saumur region. The communication provided by public institutions reflected a territorial imposition, in the sense of Claude Raffestin (1980). For example, the Centre-Val de Loire regional tourism committee promoted the castles in its territory, the castles of Chinon, Langeais, Ussé, located in Indre-et-Loire, but not that of Saumur, located in Maine-et-Loire (Pays de la Loire region). However, according to a field survey carried out, some tourists included Saumur in a tourist mobility journey starting in the eastern part of the Loire castles (Blois, Chambord sector) and Saumur, even Angers, constituted the western boundary of the destination. Other tourists associated Saumur with an influential tourist territory, forming a regional hub in its spatial practices (Piriou 2016).

To harness this "frontier" effect, a Val de Loire destination contract was signed in 2014 (and then renewed in 2018), between the French State, the General Directorate of Enterprises, Atout France, the Centre-Val de Loire regional tourism committee and regional agency of Pays de la Loire. Its ambition was to bring the two administrative regions closer together on common themes such as heritage, enotourism, Loire by bike, the art of living and tourism, and nature.

1.11. Saint-Jean-de-Monts: a community mostly gathering small entrepreneurs

Saint-Jean-de-Monts is a case featuring a fragmentation in production, divided into a large number of companies and services, on the one hand, and management provided by a public institution, on the other hand.

1.11.1. *Presentation of the place and its touristic development*

Saint-Jean-de-Monts is a seaside resort on the Vendée coast. It was established in 1867, although the development of coastal tourism had already begun in Les Sables-d'Olonne, further south, around 1823, and tourists were locally present since 1850. The Pays de Monts is a dune massif planted with conifers under the Second Empire, scattered along a sand spit oriented from the south-east to the north-east and clinging to the rocky points of Sion-sur-l'Océan[94], located to the south. The prolongation of the dune cord is the island of Noirmoutier. The municipalities of Notre-Dame-de-Monts and La Barre-de-Monts-Fromentine lie to the north of Saint-Jean. The Breton Marshes (a formerly active salt-producing area until the 18th century, later transformed by human action into agricultural marshes through desalination and land reclamation) is isolated towards the interior.

In 1867, the municipal council of Saint-Jean decided to acquire a plot of forest and dune massif belonging to the state. This created a junction between the beach and the town located inland, 3 km from the sea. Until then, as elsewhere, the beach was barely occupied by a customs cabin and frequented by fishermen who stationed their boats. There were also land pirates, who lit fires on the beach for ships to run aground and plunder them. The villagers' habitat concentrated on the town, opposite inland, and extended into hamlets and scattered dwellings along the road (the current departmental road #38), which winds between the dune massif and swamp.

The first tourist establishments were built of wood on the beach itself. From this first core, Saint-Jean's urbanization continued at a slow pace. Apart from a four-story hotel built in 1902 at the end of the *Rue de la Mer*, tourist accommodation was offered by socially heterogeneous villas. A second group, the *Plage des Demoiselles*, developed towards the south. There, a housing estate was gradually built at the junction of Saint-Jean and Saint-Hilaire-de-Riez. The creation, and subsequent abandonment of a military airfield created for the American army during World War I, opportunely ensured this new gap in the forest massif in the 1920s.

A second phase of major tourism intensification took place after World War II and after reconstruction, in the 1950s. In particular, it was mainly marked by the

94 Tourist hamlet in the municipality of Saint-Hilaire-de-Riez.

creation of an embankment promenade and the construction of buildings in towers of 8–10 stories, and five-level bars. Densification was pursued until the 1970s, creating morphological diversity in the buildings, towards the north (see, for instance, the so-called Arc-en-ciel complex), while the dune massif was preserved. Towards the south, the constructions extended over the municipality of Saint-Hilaire-de-Riez. On the latter's territory, two beach stations were developed discontinuously: *Plage des Mouettes* and *Plage des Becs*. The coastal law passed in 1986 put a halt to urbanization along the coastline.

1.11.2. *The constitution of a cluster*

The creation of a cluster in Saint-Jean-de-Monts in the mid-1980s was derived from the perception of a state of crisis by the local actors. After confirming that the beach station's trade area was shrinking and that the average spending per tourist was decreasing, the initial response came in terms of investment in infrastructure, expected to trigger the return of wealthier tourists. The opening of a golf course in 1986 (accompanied by a real estate offer in villas, and the installation of a balneotherapy center) by a regional investor provided the station with more sophisticated facilities in its northern area, something which was supposed to better meet the expectations of wealthier clienteles. Apart from large facilities, the transition from collective housing to villas reflected the evolution of the social representations of modernity. However, the image of the station did not fundamentally improve, and the effects were considered insufficient. This relative failure led, in turn, to a second response.

The reaction, developed in a second phase, came down to structuring the system of actors through a mediating organization with the status of a local semi-public company in July 1993.

> This local semi-public company became the legal support for the tourist office and ensured the operation of the nautical base, the Parée-Jésus tennis courts, the archery and the parking area for Primprenelles camper vans, while managing the Odysséa convention center (Ouest-France, June 27, 2013).

All the actions implemented by the semi-public company were based on a network of professionals gathered into clubs, who met on a regular basis. 86 professionals took part in the meetings. The semi-public company was responsible for the secretariat and coordination of those meetings, as well as the mediation between professionals, whose heterogeneity provided a diversity of professional cultures and technical languages, without omitting the tensions related to clashes between sometimes strong personalities.

This did not prevent tensions, though. On the one hand, there was a clash with the town hall services (namely, with the town planning department), regarding the direction chosen. On the other hand, problems arose among the traders who were not integrated into the semi-public company (those at the beach and those in town).

1.11.3. *The actions*

In the first place, the public actor expressed its intention to revitalize the place by initiating a rehabilitation of the coastal promenade. The latter had been built around 1950 and bordered the seafront. Its buildings were made of parallel-elongated bars, which were perpendicular to two elements located on *Place de l'Europe*. It had been designed according to the urban planning model of the time, which favored cars, with two-lane traffic and parking spaces. On the contrary, its refurbishing was imbued with new designs, which called into question the car's prominent role. Traffic was therefore limited to one lane in each direction towards the interior, as close as possible to the buildings; whereas, parking was organized into car parks. Spaces were reserved for the soft circulation of cycles in particular, of "rosalies" (four-wheeled and pedal vehicles), and for the pedestrian promenade (for which the view of the sea and the beach was reserved).

Once the semi-public company was formed, it would bring actors together and engage them in several joint actions. The first action was to define the strategic positioning of the beach station in order to distinguish it from an undifferentiated seaside supply along the French coasts. The image of a safe beach, ideal for children and family visits was finally chosen. The slogan "*let them play... you can breathe*", which targeted parents, was featured on a double page of the beach station's welcome magazine. The actors exploited the place's characteristics, in particular, the vast expanse of sand which gently extended over 8 km and was surprisingly flat. Indeed, the movement of the tides was not accompanied by the digging of dangerous tarpaulins (along the Channel coast) or *baïnes* (along the Landes coast), agitated by dangerous currents. To give credibility to this orientation, further actions were implemented: totems bearing distinctive images were deployed along the beach, so that children could find their way back and avoid getting lost. Likewise, the tourist office hosted a games corner and distributed cards with games.

Another activity for children, the "Kid's Follies", was organized during school holidays. It included shows, exhibitions and art workshops. Obtaining the children's station label strengthened communication. Events were not only chosen and organized during the summertime in order to liven up the station, but also before and after the season, so as to spread out attendance in spring and autumn, and thus improve business profitability.

Date	Event	Description
April	Roller attitude	Introduction to roller skating in dedicated and protected areas set up along the promenade
April 5th and 6th	Decorative moments	Decoration fair
April 12–17th	Kid's Wednesdays	Discovery of activities for 6–12-year-olds
April 12–27th	Kid's follies	18th children's festival, Arcimboldo theme
May 3rd	The Dune	Hike at nightfall. 8, 15 or 20 km circuits
May 1–3rd	The Breaking Wave	Street theater festival
May 10–12th	Antiques fair	7th edition
June 7–10th	Theatrical spring	Theatrical meetings for schools, colleges and high schools
June 14th and 15th	*La Mie câline basket go*	National junior tournament
June 21st and 22nd	"Naturel'Monts"	Meeting for sustainable development, habitat and organic market
June 28th and 29th	International triathlon	24th edition
July and August	"Toutain" exhibition	Over the esplanade and at the Palais des Congrès
July and August	The Breaking Wave Festival	Outdoor festival including street performances and live music. Free of charge
July and August	Kid's Wednesdays	Discovery of activities for 6–12-year-olds

Date	Event	Description
July and August	Beach planet	Discovery of beach sports: beach volleyball, beach rugby, zen mornings
August 9th and 10th	*Cinété*	Cinema screenings on giant screen on the beach. Free of charge
August 16th and 17th	Games destination	Free access games for all
September 20th and 21st	*Caval'oceane*	Equestrian competitions and activities on the beach
September 27th and 28th	*Vélocéane Du pont d'Yeu*	Lively family walks
October 10th to 12th	Creative passions	Creative leisure lounge
October 25th to November 2nd	*Ci t'as la trouille*	Cucurbit exhibitions, workshops and family rally
November 8th and 9th	Zen Atlantic	Wellness fair. Workshops, conferences and exhibition-sales
November 15th and 16th	Wine and local products fair	Gastronomy week. Workshops, conferences and exhibition-sales

Table 1.3. *List of events organized during the year 2008 by the semi-public company Saint-Jean-de-Monts Activities (welcome document published by the semi-public company, produced by the authors)*

The semi-public company also implemented actions on the theme of the environment and sustainable development. In the field of transport, an electrically powered "Emma" shuttle provided connections between remote accommodation and the beaches. In addition to the one dedicated to the environment and Emma, five other clubs brought together 80 professionals: the business club, the *e-marketing*, the restaurateurs, leisure and real estate agencies. A weekly schedule was distributed with daily information and recommendations for the general public. A negotiation with the National Forestry Office and funding agents mobilized by the community of municipalities made it possible to develop cycle tourism routes in the forest and, together with the routes established in the urban fabric, constitute a clearly signaled and secure 70 km-network that irrigated the entire territory.

A balance sheet approach can give the impression of ease, or even of certain inevitability. In reality, every action was thought through, then tested, and some of them, failed. For example, the residential leisure village project was not successful. The goal was to modernize the seafront buildings (built between 1955 and 1970), which, by the end of the 20th century, had a degraded image. This question of the gap between the evolution of mentalities and the rigidity of infrastructure particularly affected tourist places built speedily and within the framework of development operations (such as those carried out by DATAR after 1963). Contrary to this, coastlines such as the Côte d'Azur were formed historically, on a progressive basis, and renewed over time. Corrective solutions were attempted. For example, in 1991, the Balearic government adopted a law limiting the creation of new hotels to renovation and upgrading operations: for example, the construction of a three-star establishment at the place of a one-star hotel. However, this decision could be enforced due to a confrontation between a powerful public actor and a small number of hotel groups. The context was very different in Saint-Jean-de-Monts. The semi-public company faced a multitude of small owners (up to 400 for a building), who were also aging and therefore reluctant to invest. The attempt to market a proposal including accommodation and activities for the pre-season was poorly echoed by customers. In fact, the majority of tourists refused to lock themselves into packages which they perceived as rigid. Most came from France, and even more from nearby regions, which meant that they built their experience day by day.

At the beginning of the 2000s, the revitalization of the place represented a successful venture, marked by the rise in land prices and investment in the creation of tourist residences (in particular, at the regional level). The achievements of the local group then passed under the control of a national-scale company, although the transition seemed related to a strategic error. Under the control of national companies, the Lagrange Vacances group and Eurogroup were rebranded as "Madame Vacations". On a similar note, campsites experienced an upscaling which was accompanied by heavy investment, particularly in the creation of heated swimming pools and wellness facilities.

From 2012 to 2015, the semi-public company was part of the policy of the Pays de la Loire region and subscribed to a regional innovation platform project. During that period, the actors held meetings on a regular basis and consulted with each other in order to define the further development of the beach station.

1.11.4. *Evolution of the structure*

The cluster was challenged by the institutional change induced by the NOTRe law of August 7, 2015 (*Nouvelle organisation territoriale de la République*), whose provisions on inter-communality had strategic effects for tourism. This amendment was ruled upon and implemented. After three years, the director of the public company, Pascal Nerrière, contented himself with being able to hand over the "baby" to the elected officials, who would be chosen via the ballot box in March 2020.

1.11.4.1. *The event*

Until that law was passed, tourism fell under municipal competence and could only be transferred to the intermunicipal level following a favorable and explicit vote from the municipal councils. The new text immediately placed tourism at the intercommunal level from January 1, 2017. The community of municipalities, made up of five entities (Saint-Jean-de-Monts, Notre-Dame-de-Monts, La Barre-de-Monts-Fromentine, Le Perrier and Soullans) was therefore subject to the implementation of this public policy on tourism. However, among those municipalities, the first two had tourism as their main economic driver and benefited from a coastline; whereas, the three other municipalities were quite distinct. The municipality of La Barre-de-Monts comprised two different areas: the town and Fromentine. The second element was marked by a maritime culture and animated by the infrastructure connecting it to the island of Yeu. For their part, Le Perrier and Soullans were hinterland municipalities. Soullans is located on the outskirts of the town of Challans, the largest municipality in the north of the department. The profile of La Barre-de-Monts is more *maraîchin* (proper to the Vendée region) than maritime. Beyond these nuances, heterogeneity was also marked in terms of tourism because the reception capacity of Saint-Jean-de-Monts beach station exceeded that of other municipalities by far. In fact, Saint-Jean-de-Monts is the second largest French beach station, with approximately 150,000 beds, mainly in outdoor accommodation.

Tourism management shifted towards a new structure: a local public company[95], whose mission was to concentrate the actions in the field of tourism. Those relating to culture or physical activities (sailing stations, adventure parks, etc.) were separated and fell under the responsibility of municipalities. In Saint-Jean, the semi-public company bounced back and focused on those areas.

> At the beginning, the name 'pays de Saint-Jean-de-Monts' [given to the new structure] posed a problem, especially in La Barre-de-Monts. That issue was later solved because we worked on the basis of the five pillars of the five municipalities. Work on the identity of each municipality was carried out (the same which had previously been exclusively tackled by Saint-Jean-de-Monts), but the options were finally declined in terms of circuits and products. For example, in La Barre-de-Monts it was the maritime dimension and the broad horizons that were emphasized. Furthermore, Le Perrier wanted its own tourist information office, they got it. It signified an investment of €350,000, but it contributed to building the region[96].

The local public company integrated the tourist dimension of the six municipalities of the previously created community. The headquarters were located in Saint-Jean, as was the case for the community of municipalities. This change in dimension did not seem to pose any problems internally, as the tourist offices of the different member municipalities had already been working together on good terms. The public company organization chart defined seven poles (territorial influence, reception and consultancy on stays, development, prospection and support, marketing and communication, digital innovation, and finally the resources center, which was shared with the semi-public company Saint-Jean-de-Monts activities). The welcome center placed the tourist information offices of the member municipalities under joint responsibility. However, the staff numbers and system reflected the heterogeneity of the whole (five permanent staff members in Saint-Jean, three in La Barre, two in Notre-Dame; whereas, reception was only seasonal in the other two municipalities). Likewise, responsibilities were distributed. Therefore, the development, support and prospection center were piloted by the former director of the tourism office of Notre-Dame-de-Monts. A management committee brought together the heads of the divisions and the director of the public company; whereas, each department held a weekly meeting.

---

95 Local public companies (SPL, *Sociétés publiques locales*) have 100% public capital; whereas, semi-public companies (SEM, *Sociétés d'économie mixte*) are mixed economy companies which share their capital equally with private shareholders.

96 Interview with Pascal Nerrière, at the moment director of the local public company, March 4, 2019.

1.11.4.2. *Tourism at the territory's peril*

The islands of Noirmoutier and Yeu retained their autonomy, thus breaking the dynamic. The tourist fabric's continuity in the south was interrupted on an institutional level, since the municipality of Saint-Hilaire-de-Riez (whose area known as Sion had been involved in tourism since 1870) was attached to the community of municipalities of the region of Saint-Gilles-Croix-de-Vie. The latter had over 48,000 inhabitants divided into 14 municipalities, including Saint-Hilaire (the most populated, 11,000 inhabitants) in front of the port of Saint-Gilles (7,655 inhabitants), and beach stations such as Brétignolles-sur-Mer and Brem; whereas, the other components were located in the hinterland. The boundary between the two communities crosses the coastline and the so-called *Demoiselles* district, straddling the territories of Saint-Jean and Saint-Hilaire. This rupture can be appreciated in the urban landscape. The development of the embankment in these last two municipalities was not carried out at the same time, and therefore reflects the ways in which urban planning models have evolved. Formerly implemented in Saint-Jean, the separation of traffic led to a clear and physical distinction between the cycle path and the pedestrian promenade; whereas, in Saint-Hilaire, this mechanism (considered to be more accident-prone) was replaced by a mixed lane. On the other hand, revegetation symbolically reintroduced the dune landscape.

In a similar sense, in the municipality of Sallertaine (contiguous to the territory of Saint-Jean-de-Monts and Perrier), public and private dynamics undoubtedly benefited from being better integrated into the tourist centrality. However, the village fell within the community of municipalities of Challans. Without the separation being complete, the tourism office of Saint-Jean-de-Monts publicized flyers praising the visit to Sallertaine. Likewise, the Océan-Marais de Monts community was advertised on the displays of the Challans office. However, we could note that each territory tended to exclude the other. Therefore, the respective tourist maps presented them as islands and completely erased the existing connections. The tourist map of the CM of Challans Gois also presented two facets: on the front, the territory appeared disconnected from that of the neighboring CM of Océan-Marais-de-Monts (apart from a few links); whereas, on the back, the two of them appeared, as well as the road network.

These multiple splits resulted from the combination of cultivated identities (notably, insularity) and the games of politicians and their personal ambitions. Nevertheless, these centripetal forces were compensated for by the relationships established between the management of some of these structures – firstly with the CM of the region of Saint-Gilles-Croix-de-Vie. On the contrary, the Vendée des Iles union (bringing together since 1992 the municipality of Île d'Yeu, the CM of Océan-Marais de Monts and the community of Challans Gois, from where Noirmoutier left) was not remarkably active, except for the development of a tourist information website. The public company pays de Saint-Jean-de-Monts worked more with Saint-Gilles, because the two communities of municipalities (CM) shared the same coastline and employment area. There was no centralizing body. Vendée tourism did not fulfill that role. For example:

> We have historically worked with Open system, which is the reservation tool for the region and Vendée tourism, but we are going to switch to Eloa – more efficient – which is integrated into the new professional on-demand site, as it facilitates services sales for local structures. We considered maintaining marketing operations, and decided to pursue them, which in turn helped us to "attract more professionals" (interview).

The interest for the public company lay in increasing the links with the CM's tourism or economic development actors.

1.11.4.3. *The new institutional landscape*

The change brought about by the reform led to an upheaval and a landscape reorganization. Tourism became an intermunicipal competence, but culture continued to be managed by the municipalities. As a result, events organized for children (notably "Kid's Follies" in the spring, or "Ci t'as la trouille", its autumn counterpart) escaped the scope of the public company, in charge of tourism; whereas, families constituted the destination's target. The same happened with events for attracting the public such as "Planète beach", which offered summer activities in the fields of art, sports or leisure, "Caval'oceane", dedicated to equestrian arts, or the "Breaking Waves". The other municipalities equally retained control over their cultural events.

The semi-public company also maintained the reception of all business trips. However, this distribution raised some issues. The public company did not own the space, which belonged to the municipality and was managed by the semi-public company. However, the CEOs of the public company wished to have a space for local producers in the reception hall, open to private concierge services, etc. The same applied to professional mobility or sport, the semi-public company had the infrastructure while the public company was the business provider. A person in charge of coordination was missing. The public company also had an authorization for marketing products. The change in the institutional level therefore induced a discontinuity between the management bodies responsible for tourism and those responsible for culture. By remaining managed at the municipal level, culture embodied the last means for demonstrating autonomy.

Partnerships were established with other structures. Receptive activities for groups were transferred to Vendée tourism, the departmental institution. It was officially committed to promoting the Saint-Jean-de-Monts brand among leisure groups. For the adult group and all groups of children, agreements were concluded with *Objectif séjour* and *D'click tours passion*, the two receptive centers in Vendée that specialize in school and youth trips.

Relations with the services of the Océan-Marais de Monts community of municipalities (CM) were close, and the two structures were coordinated. Indeed, territorial influence is not limited to tourism. Other issues, notably residentialization (the long-term primary or secondary residence of new inhabitants), were also concerned. However, this led to problems with the development of residential spaces as well as business zones. The workshops for businesses were shared between the public company and the CM, because the problems and challenges were the same. This implied relations between the public company's territorial influence center and the CM'S culture and communication department, in particular, through a strategic committee, in which the director of the public company was also involved (devoting 10% of their activity to the CM's tourism direction). The CM also managed five cultural and tourist establishments (Table 1.4). Support for professionals was provided in partnership with the Chamber of Commerce and Industry of Vendée. The environmental question was also coordinated between the public company and the CM. Both structures were based on the approach to corporate social responsibility initiated at the departmental level.

Name	Municipality	Description
Le Daviau	La Barre-de-Monts-Fromentine	Eco-museum of the Breton Marshes
Biotopia	Notre-Dame-de-Monts	Eco-museum of the sea and the forest
Kulmino	Notre-Dame-de-Monts	Water tower converted into a viewpoint
Déambul	Le Perrier	Skiff and carriage ride in the marsh
Charles Milcendeau Museum	Soullans	Former residence of the painter

Table 1.4. *Cultural and tourist establishments managed by the Océan-Marais de Monts community of municipalities. Created by the authors*

The relations between municipalities regarding tourism needed to be built. A discussion was initiated in 2017 to articulate the two levels under a unifying brand but respecting the positioning of each territory, due to the different identities which have been already mentioned. The common brand was adopted: the region of Saint-Jean-de-Monts established itself following a reflection carried out jointly through workshops held in 2018 and 2019. A total of 100 professionals and ambassadors were committed to implementing specifications. The interest for the actors was to be marketed by a local, close structure, something that implied a far lower commission, 3%, than those charged by the tourism offices, often higher than 15%. Furthermore, the relatively low share of foreign customers (10%) made room for this choice.

1.11.4.4. *The adaptation of professionals*

For professionals, the hurdle was also difficult to overcome. Within the semi-public company, before the transition to inter-communality, in Saint-Jean-de-Monts the 86 professionals who took part in clubs intended to increase the quality of services. On the one hand, it was necessary to manage the change in dimension. The period of development of the new organization made it possible to open the game to new professionals who had not previously been involved, or even those who were not directly related to tourism. Workshops were set up with a vote on decisions during sessions conducted by a consultant. The reflection was presented to the board of directors. The transition from the semi-public company to the public company status had the effect of excluding professionals from decision-making, since the capital of a public company is 100% owned by public actors.

A technical committee was organized to reintroduce them, but in a consultative form. It was voluntarily limited to 25 members to be effective and open to personalities representative of vertical networks, because "they have other prisms than the Monto-Montois prism". Four meetings were held in two and a half years. The last edition also brought together the board of directors, the management of the public company, as well as the general director of community services. This assembly approved a proposal called "Ambition for the future", a common position which had to be validated by the elected officials during the following elections (March 2020). The proposal came down to "acting as an engaging sustainable destination […], that is to say, an attractive spot where everyone was involved"[97].

On the other hand, the clubs also evolved. First, a distinction was made between clubs and sectors. In some cases, the public company actors had a coordinating role in the service's upscaling, and in the strategy for earning and building loyalty. This intervention was redefined by the distinction between three levels: beginner,

97 Excerpt from the interview with Pascal Nerrière, March 4, 2020.

confirmed and expert, which acknowledged the heterogeneity of professionals. In other cases, professionals required less support. Diversification also took place. In catering, a special operation around gastronomy was carried out: dishes and identity products appeared on the menus, such as eel, market garden flan, cuttlefish rata, without disrupting the region's cuisine. Finally, three themes received particular attention: the environment, an outstanding welcome service and everything related to the customer experience. Pascal Nerrière, director of the public company, insisted: "we also want to play on the proximity, the interaction between the professionals", because the transition from the semi-public company to the public company gave the professionals the feeling of being abandoned.

During the delivery period of the new structure, further actions were pursued. The new website was prepared and relaunched. Trilingual, it enabled the tourist office to send applications to reach a first-class category. Actions around sport continued: the *trail*[98] station label was obtained and inaugurated on September 1, 2019. It was the first of its kind on the Atlantic coast, aiming to modify the image of Saint-Jean-de-Monts (still perceived as "too popular" for part of the media), and evolution was noticeable in many other aspects. Activities made considerable progress. The tourist tax increased from 10 to 12% after 2017 to reach 1.2 million for inter-communality (including 70% in Saint-Jean-de-Monts), an increase of 7%, and attendance was estimated to 6.5 million overnight stays that year[99].

Reflection was pursued. The concierge service was among the topics discussed. The challenge was to optimize the time spent by tourists by freeing them from the constraints which held them back. It became necessary to carry out an inventory of needs, which were quite heterogeneous. For example, in La Barre-de-Monts, 600,000 people travel every day to reach the island of Yeu. The intention was to free them from their baggage so that they could further enjoy their passage. Another project was based on the question of the tourist office of the future, transforming it into a meeting place with *La Folie*, the place of attractiveness. Businesses are welcome, and the tourism office attracts tourists.

98 "The *Trail®* Station concept. A *trail®* station is a place 100% dedicated to *trailing*, which offers courses, services and tools both for beginners wishing to learn and be supervised, as well as for enthusiasts looking for ways to organize courses, or complete weekends in one place. Each *trail®* station is attached to a region eager to help you discover its most beautiful landscapes and its best routes and to welcome you for a day, a weekend, or a week". Available at: https://stationdetrail.com/fr/ [Accessed March 4, 2020].

99 Estimated according to Flux vision orange.

1.12. Innovating tourism: the reason for being of the "Tourism InnovationLab"[100] cluster

Tourism InnovationLab was designed with the goal of developing a cluster, bringing together not only institutions, businesses or start-ups, but also training centers (ESTHUA, ESSCA, CCI, etc.) and research laboratories within the Pays de la Loire. It saw the light of day in 2015 and its main missions are: the coordination of a network around tourism innovation and support for project leaders and start-ups.

1.12.1. *Initial system and intentions of the Pays de la Loire region*

In 2015, the Pays de la Loire region set up RTI (Research–Training–Innovation) mechanisms in the territory's promising sectors. Included in the 2014–2020 regional plan for higher education, research and innovation, these RTI action plans aimed at making the skills and know-how of regional actors visible, by promoting cooperation. Inspired in a principle close to that of competitiveness clusters, the intention was to combine academic research with the socio-economic world, to develop regional research, to promote scientific skills for the benefit of regional development and to disseminate scientific culture.

The Pays de la Loire region, comprising five departments and nearly 3.8 million inhabitants, developed around a dozen RTIs on its territory, following a territorial diagnosis.

Relying on specific university capacities (having the largest European training center in tourism with ESTHUA Tourism and Culture, as well as the UFR (Unité de Formation et de Recherche) of the University of Angers gathering over 3,000 students and 25 teachers-researchers specializing in tourism and heritage), the Pays de la Loire region also displays an authentic regional tourism dynamic (sixth largest French destination, 43,000 jobs, 8% of regional GDP) with varied resources: Nantes (cultural city), the Vendée coast, Puy du Fou, the 24 Hours of Le Mans, Saumur, etc. The RTI tourism, called Angers TourismLab, was launched on April 10, 2015, for a period of five years. Supported by the University of Angers, it had 12 founding members and 11 associate members, including the three regional universities, ESSCA, the communities, Accor hotels, Unimev, Fontevraud Abbey, *Voyage à Nantes*, Saint-Jean-de-Monts station, Appart'city, Angers technopole, etc.

100 This case was written by Ms. Aude Ducroquet, lecturer, University of Angers, who also took part in exchanges and field activities, notably in Amnéville, Center Parcs and the Louvre-Lens.

The actions of Angers TourismLab encouraged the development of structuring projects, among which we can mention the following:

– Research: the formation of a Scientific Interest Group (SIG tourism studies), reuniting 150 researchers from various disciplines from 34 laboratories and 17 establishments in the Greater West (from Caen to Pau). Supported by the CNRS, it was officially launched in May 2019. The RTI tourism also enabled the financing of 21 collaborative research projects and the recruitment of doctoral students and postdoctoral students.

– Training: the University of Angers, the Chamber of Commerce and Industry of Maine-et-Loire and the Higher School of Agriculture (ESA) joined their forces to form a school of gastronomy, wines and spirits called the Gastronomy Campus, offering over 36 training courses. In addition, in August 2018, the certification for the campus of tourism, catering and international professions and qualifications of Pays de la Loire was issued. It brought together all professional sectors in the tourism sector, was jointly supported by the Rectorate of Nantes and the region of Pays de la Loire and partnered a network of actors in a collaborative dynamic to develop an initial and continuing training supply related to the tourism sectors (ranging from CAP level to doctorate).

– Internationally, several partnerships were established, whether in terms of training, research or mobility: China, Mexico, Canada, Indonesia, etc. Furthermore, the Tourism Summer Schools granted training during July and combined it with a discovery of the regional territory.

– In terms of influence, the International Tourism Festival (FIT) was created in 2017. The event took place over several days and involved different manifestations: scientific seminars, conferences, exhibitions, concerts, book fair, etc.

– Innovation: the creation of a cluster, Tourism InnovationLab (TIL), made it possible to bring together over 50 members coordinated by the collective (regular thematic meetings, an annual day, etc.), supporting start-ups (over 130 regional start-ups) via training workshops and a call for projects (31 project leaders received assistance in training, incubation and experimentation). The TIL was part of the national network of France Tourism Lab incubators, led by the General Directorate for Enterprises (GDE). It is this last mechanism, which is particularly relevant to this section, and will be presented and analyzed in the upcoming lines. Aude Ducroquet co-chaired the TIL, along with Thomas Yung (entrepreneur), Jean-Jacques Micoud (LTS Saumur) and Benoit Lequément (Sarthe tourism). As the researcher was at the heart of the system, this engineering research position (Allard-Poési and Maréchal 2014) granted the close observation of the inner workings of this LTS.

1.12.2. *The birth of the Tourism InnovationLab cluster*

The implementation of this cluster was part of the highly specific context of the RTI and stemmed from the regional desire to make tourism a center of excellence.

1.12.2.1. *Getting prepared for the cluster through a feasibility study*[101]

In 2015, the Pays de la Loire regional agency launched an economic feasibility study with a view to setting up an operational provisional structure for Tourism InnovationLab. This study was carried out by an independent consultancy firm. With that goal in mind, a benchmark, as well as interviews with institutional and private structures, were carried out.

The purpose of this study was:

– to define the possible strategic positioning, expectations from regional actors and scope of action;

– to develop recommendations for operational provisions (such as the legal framework, the technical and economic dimensions);

– to develop a roadmap by identifying development levers, prioritizing actions, defining indicators for measuring results and estimating the human and material needs required for optimal functioning.

Working groups were formed and met from March to June 2015.

The first workshop, which took place on March 27, 2015, gathered 37 participants, including 19 companies and clusters, as well as 18 institutional and academic representatives. Its purpose was to identify issues and problems, and to list expectations and needs.

The second one, dating from April 17, 2015, brought together 25 participants, including 15 companies and clusters, and 11 institutional and academic representatives. It helped establish the strategic positioning of the TIL, identify professions and missions, and define its supply.

The last two workshops involved 18 participants, with approximately 70% companies and clusters and 30% institutional and academic representatives. The third one, which took place on May 22nd, focused on the structuring of the action plan and the definition of the economic model. The last one, held on June 23rd, outlined the schedule and developed the operational roadmap.

---

101 This information has been drawn from internal documents and archives in our possession.

The consultancy firm also carried out around 10 interviews with the managers of the three regional innovation platforms, the director of RTI tourism and six companies and professional federations.

A presentation to key actors and to Angers TourismLab members took place in September 2015. The major tourism trends were revealed. The consultancy firm also emphasized that "the tourism sector still had a very limited culture of innovation at the moment. By means of an incremental approach, the tourism sector had nonetheless been able to gradually renew its supply and to adapt to new consumption trends"[102].

Three strategic goals were defined by the consultancy firm:

1) to unite, animate and raise awareness;

2) to encourage the emergence of, support and experiment with innovative projects;

3) to contribute to skills development.

1.12.2.2. *Organization of the cluster and its pilot committee*

The TIL cluster was launched on November 18, 2015 in Angers with the purpose of "promoting the sharing of experiences, encouraging the emergence and supporting new innovative projects"[103].

The provisional association gathered 26 companies. The two co-presidents were Thomas Yung, entrepreneur and specialist in Hotel web marketing, and Jean-René Camus, representative of the Chamber of Commerce and Industry of Maine-et-Loire, and manager of a hotel in Saumur. The bureau also included a representative of the Regional Innovation Platform (RIP) for Innovation and Design of the Visitor Experience (IDVE) from Fontevraud, and a representative from Loire-Atlantique Development.

This provisional association held meetings approximately every two months to make progress in governance, communication tools, the definition of missions and goals, etc. These sessions were sometimes supervised by external professionals: communications agency, creativity *coach*, the director of France cluster, etc.

102 Economic feasibility study outlining the operational provisions for the TIL within an RTI tourism center of excellence in Pays de la Loire.

103 See: https://www.univ-angers.fr/fr/universite/actualites/tourism-innovation-lab.html (Accessed March 17, 2020).

The governance problem could not be resolved. The tensions between the already existing institutional actors were palpable. The RTI touched on the sensitive area of innovation, which was increasingly beginning to appear at the national level, supported by the General Directorate for Enterprises. Even though the region financed the RTI, the economic development agency was not ready to lose its *leadership*. It therefore opposed the TIL's associative status, and it was impossible to achieve synergies at the workshops on tourism innovation. Sometimes meetings were even held in parallel with the TIL workshops on the shared topic of tourism innovation.

It was decided that (or rather, without consensus) the TIL would be managed by the RTI innovation commission and coordination would be granted by the innovation project manager. Therefore, the pilot committee joined that of the RTI, and strategic decisions were made within operational committees and pilot committees, mostly led by the different funding organizations.

For even more operational aspects (missions, communication, coordination, etc.), the members quickly agreed on the proposals from Solène Chesnel, the TIL innovation mission manager.

Nearly two years were required to move from the provisional association to the TIL itself.

1.12.2.3. *Coordination of thematic workshops and days dedicated to innovation*

The life of the TIL was organized, and thematic workshops (TIL meetings) were implemented regularly at the regional level. We can cite, for example:

- workshop on design thinking and CTI/ITH aid in Fontevraud;
- discovery of virtual reality and augmented reality in Laval;
- the influence of the digital on tourism in Angers;
- enotourism in Saumur;
- acoustics at Le Mans;
- gourmet tourism in Angers;
- teleworking in Nantes;
- territorial innovation in Sables-d'Olonne;
- the awakening of the senses through sound at Le Mans.

At each workshop, local tourism and innovation actors were invited to properly introduce their activities and discuss the possibility of working together.

The Innovation Days were renamed the Tourist Innovation Meetings (TIM), then the Tourism Spring, and organized on an annual basis. Three editions took place during the International Tourism Festival (ITF) in Angers, and the last one was carried out in collaboration with Anjou tourism and the campus of professions and qualifications:

– June 2017: the theme was innovation in tourist offices and the regional supply to meet such demand.

– March 2018: in partnership with the Transat chair and Montreal's MT lab, start-ups and guests raised the following issue: What innovation, what practices can be proposed across the Atlantic when the city becomes a playing field?

– March 2019: focused on the theme of emotions and feelings at the heart of the tourist experience.

– March 2022: discussed questions related to the sustainability of tourism.

– March 2023: on the metamorphosis of tourism.

1.12.2.4. *Support for project leaders and start-ups*

A particular type of audience quickly appeared during these meetings and workshops: project leaders and start-up entrepreneurs. The first to join the network acted as driving forces and helped define the place of TIL in the entrepreneurial landscape: Baludik, Akken, Sumwhere, etc. Supporting these young companies swiftly became TIL's main mission. During each thematic encounter, a specific workshop was organized for start-ups so that project leaders could exchange ideas with each other. During TIMs, demonstration spaces gave start-ups the opportunity to present their supply to professionals.

In 2018, the first call for projects was launched in partnership with the technology park. A dozen files were received. After examination by the jury, seven project leaders were supported in terms of training, incubation and experimentation. In total, 31 project leaders were supported.

Training days are also organized: sales pitching, negotiation, social networks, legislation, etc.

Since 2018, the TIL has worked under the France Tourism Lab, a seven-member national network of tourism incubators and accelerators aiming to spread the spirit of innovation, to promote initiatives and to strengthen the visibility of innovative companies. The interest of this network for the TIL and the meetings with other members is to gain expertise and obtain national visibility. *Learning expeditions* were organized in Montreal in 2019, and at the CES in 2020.

The TIL gave these start-ups the possibility of pooling spaces at national trade fairs, such as the "Into days" held in Cannes in January 2019, which brought together all the players in French tourism innovation during three days and offered a demonstration space for start-ups, conferences, *keynotes*, etc.

1.12.3. *The place of the cluster in the Loire landscape*

The RTI system made it possible to develop numerous regional projects. However, it was not easy to federate local actors. Naming the RTI after the city (Angers TourismLab) did not facilitate regional dynamics. In addition, the political change in the region in 2015 blocked the development of the system for a few months, which led to numerous restructurings. January 1, 2018 saw the creation of the public company Angers Loire Tourism Expo Congress, a merger between Angers Loire Tourism and Angers Expo Congress. The region also saw the end of the three regional innovation platforms (RIPs) launched at the end of the 2000s.

1.12.3.1. *The cluster: a top-down desire*

Challenges did not cease to arise, and obstacles sometimes proved too difficult to remove: the system was born from a political will (a top-down approach), with a somewhat excessive ambition to turn the Pays de la Loire region into an international hub for tourism innovation. Reality on the field was far from that desire. Tourist structures were largely disparate; for the most part, small and medium-sized enterprises which were not enrolled in an innovation dynamic. Before this RTI mechanism, the region had launched Regional Innovation Platforms (RIPs)[104]. Three of these were dedicated to tourism: the Saint-Jean-de-Monts Regional Innovation Platform (RIP), a futuristic station which focused on innovation *marketing,* the IDVE RIP, interested in the visitor experience, and the underground world RIP (rethinking the Saumur troglodyte world). Criticism of the economic model of these Regional Innovation Platforms (RIP) had been the focus of the electoral campaign of the regional team elected later on. Innovation-related expenses (a multitude of RIPs and launching of RTIs) were judged too expensive for a return on investment perceived as too uncertain and too distant. Consequently, the RIP

104 Unlike the RTI units, driven by the Directorate of Higher Education, RIPs were driven by economic development.

quickly disappeared after the regional elections. The RTIs survived after they were recently launched for a period of five years. Almost two years had to elapse before obtaining a budget line for innovation. Under these circumstances, it became gruesome to launch a cluster on tourism innovation, especially considering it should not be a continuation of the RIPs. Messages were relentlessly contradictory. Moreover, as with every change of political side, the actors who had been involved in the provisions for TIL disappeared, one after the other. The elected officials' awareness on the importance of innovation in tourism came too late and only as a reaction to the various national systems that were emerging. The soil usually present when launching a cluster was not there.

To meet the needs of the field and avoid treading on the borders of existing mechanisms in terms of tourism innovation, two choices were made:

– The first was to help young projects. Even then, it was not easy to find a real position regarding the proliferation of existing structures: the territory was full of incubators, support structures, support systems; and the CCI, founding member of the RTI, was a key player in the territory. Start-ups regretted this regional layer cake and the impossibility of drawing a precise map from it[105]. In addition, specific support to tourism had its advantages, as well as its disadvantages. Startups will express this need to exchange with other start-ups who, even though they are in another sector of activity, are at the same level of development and encounter the same problems. Another dilemma: should they help every young start-up to access the tourism sector, or only provide support to innovative structures? The question of what tourism innovation is (and is not), was raised on several occasions.

– The second one involved popularizing academic research work. A newsletter (u'TIL info) presented research work on a given theme and discussed some examples inspired in progress monitoring. Although the professionals found the initiative interesting, the topics discussed were disconnected from their immediate concerns. The "popularized" articles were drawn from the *Mondes du Tourisme* journal but did not offer practical solutions to the problems encountered in the field.

1.12.3.2. *The cluster: a missing and useful link*

Despite its difficult beginnings, the TIL found its place in the regional landscape.

It met the needs of regional start-ups and was useful for bonding and breaking the isolation of the creator, "encouraged by people who had already been there"[106]. It

105 Interview conducted in 2018 with a start-up whose project proposed a specialized application to sporting practice.

106 Interview conducted in 2018 with a start-up specializing in activity rating.

offered guidance and the chance of being directed towards the right people. The notion of network also enabled entrepreneurs to get to know each other better, to mix their product supplies (example: equipment for sound walks with an umbrella and content based on the place's history) and to gain insights from other sectors "having knowledge on tourism"[107]. "As we knew nothing about it, we tried to integrate a structure of people who knew about it".

The TIL workshops and days, enabled actors from different ecosystems (innovation, tourism, support) to get to know each other, to exchange ideas, to get feedback, exchange good practices and discover innovation. The round tables offered during TIL meetings were immensely popular: "it's really interesting and enriching"[108].

The call for ITC (Innovative Tourism Companies) projects marked a notable step forward in support. It offered a more pragmatic response and a direct and privileged link with regional technology parks to obtain personalized monitoring, with sites and institutions for experiments. This was done in tourist offices (such as Ancenis), in sites such as Fontevraud Abbey, in certain towns (such as Angers), etc.

A working group was also set up with several start-ups to propose a new supply for the troglodytes. This group was suggested by the Saumur tourist office.

Participation in the France Tourism Lab, as well as the national visibility provided by the presence at various events (participation in round tables, workshops or trade fairs), were also significant elements in this dynamic. Following this concern for visibility, the presence on social networks (Facebook, Twitter, LinkedIn) was strengthened, broadcasting activities and news related to the cluster and its actors.

1.12.4. *The birth of a new cluster: a difficult positioning in an already complex territorialized system*

The Pays de la Loire initiative, the RTI, is interesting and is inscribed in the regions' desire to define their fields of expertise and strengthen them. Early on, the Pays de la Loire focused on innovation and saw it as an economic springboard: many resources were allocated in that direction and numerous systems implemented. Many…maybe too many.

107 Interview conducted in 2018 with a community platform start-up.

108 Interview conducted in 2019 during the TIL meeting in Saumur with an actor in the enotourism sector.

1.12.4.1. *A topic too distant from the field?*

The multiplication of regional, national, and local networks and mechanisms led to a proliferation of interviews, meetings, workshops, etc. Professionals, whether in the private or public sector, were unable to keep up the pace and respond positively to all requests.

For some of these, engaging in a network was synonymous with visibility and opportunities. Many took part in these systems as an imitative strategy, to obtain legitimacy or to access the organizational field (here we find the arguments of the neo-institutional theory defended by DiMaggio and Powell in 2004). However, they all defended their interests, seeking prompt, if not immediate, results.

The RTI tourism system was an attractive mechanism, and the members were genuinely interested in it (often at the beginning: legitimacy, mimicry, visibility, etc.). However, the immaturity of the mechanism hindered good intentions and the socio-economic actors hoped for concrete solutions to their current problems. They did not necessarily have the vocation to build the system. In addition, the construction phase took too long, with great uncertainty regarding the regional elections.

The same happened to the TIL, the cluster developed by the region to bring together actors involved in tourism innovation. At first, it was seen as an intruder in the landscape of regional tourism players – furthermore, as a limited-time intruder. The RIPs already played this problem-solving role, coordinating the community. Common sense would have beckoned the TIL to become a continuation of the RIPs, but this was not possible. Some RIPs were organized in project mode, and when they disappeared, these projects were taken over by institutions (this was the case of the underground worlds' RIP). Others, whose mission was to coordinate policies, disappeared, as did their dynamics.

Unlike other initiatives (for example, other tourism clusters or destination contracts), the vocation of the TIL was not to develop and sell the Pays de la Loire destination or new regional supplies, but to delineate the tourism of the future. While this ambitious goal was respectable, it did not have any immediate impact and the small structures in the Loire tourist landscape did not feel mirrored by this proposal. Similarly, tourist institutions failed to be recognized as the most innovative players (the delay in digital technology was quite eloquent in this sense). Innovation was not a flagship subject, a priority subject for the actors.

The fact of being unable to concretely address structural problems, contribute to commercial development or even provide relevant information (toolboxes, news, observatory, etc.) slowed down the initial dynamic.

This delicate positioning meant that the main goal (defining the tourist practices of the future) could not really be pursued, as there was no collective intelligence on the subject. While the deployment of start-ups made it possible to provide some answers, this was done without a global vision on the subject and lacked perspective.

1.12.4.2. *An underexploited strength*

Compared to all other existing territorial systems, the strength and originality of this cluster undoubtedly lay in the strong link between academic research and training in this sector. The meeting between researchers and tourist structures was not at the center of the TIL. However, during the annual meetings, the researchers shared their viewpoint on the central topic of the day. Few projects emerged within the TIL. On the contrary, the research projects supported by RTI tourism always mobilized socio-economic actors and great collaborations emerged, but these projects were research projects and not innovation ones. Here lies one of the weak points of a subject which was never addressed within the TIL: what do we call tourism innovation and what types of innovation should we focus on? The innovation supported by the TIL came from start-ups, most of which offered a new tool that could be used for visits, experiences, professionals, tourists, etc.

The connections with training continued to develop. Targeted training was offered in the form of seminars (commercial negotiation, legal statuses, social security contributions, social networks, etc.). In addition, a university diploma was created. It comprised training blocks accessible to all, either in modular form or in diploma form. These modules enabled the student to become familiar with the tourism sector and its challenges, business creation, communication, and innovation in all its forms (creativity, experiential marketing, design thinking, scenography, tourism, and trends, etc.).

Links with students were also developed by proposing student projects to companies, offering internships, etc.

The proximity to ESTHUA, a tourism training center, and the ESO and GRANEM research laboratories, could turn the TIL into a knowledge and collective intelligence hub, a platform for reflection on the tourism of the future, its challenges and establish a trend observatory. “A link between research, businesses, students and citizens, university is at the heart of the system for producing and transmitting knowledge” (Audet and Malouin 1986; Pinte 2006). The university’s mission is to “generate, transmit and capitalize on knowledge” (Nonaka and Takeuchi 1995); it is within that setting that “knowledge is developed and popularized” (Pinte 2006).

The experiments carried out by start-ups on the sites contributed to understanding an evolution of practices, which could eventually become widespread. In short, the TIL could act as a laboratory with the help of the entire territory.

1.12.4.3. *The system's limits*

However, the TIL was not a tourist cluster in the sense of the different definitions; its intention was not to develop the destination. The system set up in this defined territory aimed to turn the Pays de la Loire into a laboratory and develop a collective intelligence dynamic to rethink tourism and imagine the tourism of tomorrow.

Tourism InnovationLab ended in March 2023 and did not gain new support at the regional level. Support for start-ups was reorganized under existing bodies: Solutions&co and regional technology parks. Network management was taken up in a new regional body dedicated to tourism innovation: RDI tourism.

1.13. Nantes: from a little-frequented city to tourist destination

The capital city of the Pays de la Loire region has not historically been touristy, or at least, tourism has not been particularly intensive at first sight (Violier 1998a). Until the 1980s, only Paris was touristy in France. This was to such an extent that Georges Cazes – one of the academic specialists in tourism at the time, together with Françoise Potier – wrote a book published in 1996 signaling the emergence of "urban tourism", even though Paris had been visited by tourists since at least 1820. However, as in France only the capital city was worthy of it, there was no such thing as urban tourism, but tourism in Paris. In France, tourism spread towards the smaller cities with decentralization, which led major politicians to take over municipalities and to engage tourism as a lever for economic development and as a means of communication to strengthen the place's attractiveness. Despite appearances, tourism became a major issue in the construction of regional metropolises, and relevant actors dreamed of a European dimension. Martine Aubry in Lille, Alain Juppé in Bordeaux, Jean-Marc Ayrault in Nantes, etc. understood that heritage does not suffice (ultimately, all towns in France have a castle and a cathedral), but the strategy must enrich the supply while the town encourages the tourist practice of discovery; all of this involves the creation of tailor-made events and establishments.

1.13.1. *A late start for tourism*

In the 1960s and 1970s, Nantes was a city that looked elsewhere, towards the sea, and was battered by the collapse of its economic model. Port activities were shifting towards Saint-Nazaire. The closure of the shipyards in 1987 ended the city's maritime adventure. Although it inherited the function as a regional capital (first within the framework of the seventh plan, the Rectorate being established in 1967 and then during decentralization), the city was absent from tourist itineraries.

1.13.1.1. *New mayor, new outlook*

For Nantes, the election of Jean-Marc Ayrault as town hall in 1989 marked the beginning of a first and real tourism policy. From this date, culture has incarnated one of the main dynamics and components of the Nantes urban policy. This transformation began in 1990 with the installation of the *Royal de luxe* company in Nantes, specializing in street arts and featuring the fantastic giants. This troupe came from Toulouse, where it was no longer supported. Also in 1990, Jean Blaise, associate of Jean-Marc Ayrault since 1982, then mayor of the neighboring town of Saint-Herblain, launched a new cultural program in Nantes: the *Allumées* festival, investing in the city's industrial wastelands in an offbeat manner. In 1995, the city supported another now major event, the musical show *La Folle Journée*, launched by René Martin: nearly 300 classical music concerts given in one weekend at the Nantes Congress Center. In 2000, the 8,000 square meters of the former LU biscuit factory were turned into a cultural venue, the *Lieu Unique*, bringing together a national stage, exhibition venues, restaurants, a bookstore, a nursery and recently, a hammam: nearly 500,000 visits recorded in 2008. In 2007, yet another innovation came about. Jean Blaise brought the art world out from the *Lieu Unique* with the creation of a contemporary art biennial between Nantes and Saint-Nazaire, called "Estuaire" (as the town bearing the same name). It reunited around 40 artists of international stature who set up sculptures and other installations along the Loire, some of which were to remain on site. That same year, the castle reopened after a major renovation, which had taken 15 years of work and five years of total closure. The new organization granted free access to the courtyard and the ramparts, the installation of the Nantes History Museum, and almost total accessibility to four forms of impairment. In addition to this (also in the summer of 2007), under the leadership of François Delarozière and Pierre Orefice, former members of *Royal de luxe*, there was the extravagant *Machines de l'île* project. Initially, two elements were installed: the Galerie des Machines and its famous elephant. 12 m high, it carried around 30 passengers embarked in a basket on its back for a walk of approximately 20 minutes. The walk started at the gigantic naves of the former Dubigeon shipyards, revamped for the occasion. The effect was breathtaking and its success, immediate. Other creations followed, such as the Marine Worlds or the Heron Tree, with equally disproportionate dimensions, exceeding 20 m in height.

1.13.1.2. *From the ephemeral to the lasting*

Essentially, the cultural policy of Nantes was based on event organization. Nearly 20 years later, the desire to transform this one-off cultural promotion into a lever for permanent development became crucial. This was the case of *Machines de l'île* and its articulated elephant, whose inspiration stemmed directly from the *Royal de luxe* shows. The principle was to develop a sort of "theme park" open to the city and strongly attached to the imagination of *Royal de luxe* in a former port district of the island of Nantes under renovation. The *Machines de l'île* project was presented as a novelty, its intention being to turn an experience hitherto exclusively associated with events and hosted by the *Royal de luxe* troupe, into a permanent tourist establishment. That was also the case for the *Estuaire* biennial, whose a priori ephemeral character was transformed into a lasting operation in 2011, thanks to the works recovered from former biennials. A marked tourist route along the Loire helped the visitor to discover them: *Les Anneaux* by Daniel Buren at the end of the island of Nantes and the soft boat *Misconceivable* by Erwin Wurm at the Martinière canal. However, this subject was also a bone of contention. Conflicts arose, for example, regarding the Martinière canal, where a French building architect worked to ensure that the contemporary creation by Erwin Wurm – despite being intended to remain – disappeared as quickly as possible from the landscape. This was done under the pretext that it distorted the canal no longer in use, but whose imagery was too romantic to accommodate a contemporary installation. Through the *Estuaire* operation, the strategy displayed by the cities of Nantes and Saint-Nazaire sought to create a future Loire metropolitan area, with a European influence over nearly a million inhabitants. As time passed, the transition from cultural event to permanent tourist itinerary reflected an acknowledgment of the territory and the events policy conducted until then.

Two elements were characteristic of this development strategy combining culture and tourism. Since its conception, the event had a popular dimension and sought to respect the goals of artistic creation and cultural requirements. Based in Nantes, Jean Blaise defended the idea of high-quality culture, but remaining open to as many people as possible. However, it was key to transforming an essentially ephemeral event into a permanent lever for development. In Nantes, this transformation was achieved through the transition from street theater performance to theme park.

1.13.2. *The construction of a localized system of actors*

This type of construction results from the interaction between the actors engaged in a project.

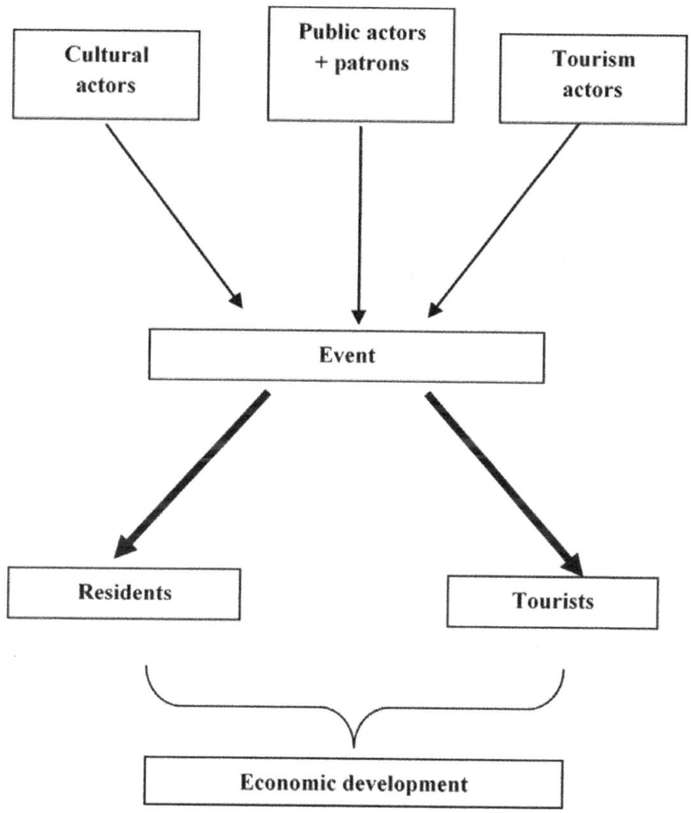

Figure 1.11. *The joint organization of an event by culture and tourism actors induces economic benefits, provided that the expected attendance from permanent residents and tourists is reached. Created by the authors*

1.13.2.1. *The project: seizing the event as a lever for development*

Fundamentally, for the cultural event to transform into a driving force with an impact on others, it must attract external populations, tourists, who, as a result of their passage, generate an economic transfer towards the city hosting the event (accommodation, catering and shops in general; Figure 1.11). Furthermore, tourists appreciate taking part in an event primarily attended by the locals. This is experienced as a guarantee of the desired encounter with the inhabitants and a proof of "authenticity". If the event only brings together residents (or mostly local inhabitants), the multiplier effect is reduced, and the activity cannot fulfill its driving role. The challenge is therefore to welcome temporary inhabitants, that is, foreigners to the place or at least obliged to settle temporarily where the event takes place, in order to generate economic benefits for the territory. The festive, as well as popular, dimensions of event policies always have a priority: urban wandering and the metamorphosis of public spaces. The crowds who concentrate along the elephant's route are reminiscent of old-time large gatherings in this part of the city, where people used to assist in the launching of newly built ships. The central question therefore concerns the links that the actors of culture and tourism can weave between them in order to enhance the conditions for tourist attendance. In general terms, this relationship is difficult in France. Cultural circles despise those involved in tourism. The feeling is mutual. Those involved in tourism criticize cultural actors for the extreme secrecy of their productions and subsidization, deemed as excessive for cultural events. The actors of culture mock the commercial dimension of tourist activities and so-called "mass", that is, popular, tourism. Figure 1.11 models the conditions for transforming a cultural event into a lever for economic development.

1.13.2.2. *Chaotic beginnings*

At the beginning, the system in Nantes did not work at its best. Cultural managers had a vision of culture oriented towards the general public (and therefore rather favorable to tourism), but the connection with tourism actors was not smooth. Relationships between tourism and culture stakeholders were tense. Before 2012, the tourist office favored business tourism in its strategy and relied on conference customers. The director considered business tourism as the driving force for attendance to the city and did not believe in the capacity of cultural policy to fulfill this role. In addition, the cultural policy of Nantes was driven by strong, famous personalities who were invited (and imposed on Nantes) by the mayor. Consultation and the idea of networking did not take hold between people, probably because the integration of the different actors was not carefully thought out and no one was really working to bring together the actors from the two environments. Therefore, two organizations related to tourism development in the city developed in parallel. The *Machines de l'île* of Nantes, as well as the castle of Nantes, were managed and promoted by a mixed economy company (the semi-public company *Nantes Culture Patrimoine*), while the tourist office (with associative status) and the Congress center (semi-public company) worked alongside them. For its part, the *Estuaire* biennial was managed by the *Lieu Unique* (Morice and Violier 2009). The way in which the projects were characterized by the different actors reinforced the differing perspectives. The case of the *Machines de l'île* is quite revealing in this respect. The facility can be compared to the concept of "theme park", not ex nihilo on the urban outskirts as usual, but on the contrary, within the city's ramparts. What made its uniqueness was the setting at the very heart of the city, pretty much as a fun fair. However, the facility was not perceived and presented as such by the project's actors and masterminds, rejecting the very idea of a "theme park", but rather emphasizing the concept of "urban theater". In a certain way, this opposition demonstrates the still existing gap between tourism and culture in Nantes. What the actors of culture feared the most was being reduced to simple and vulgar producers of entertainment parks, synonymous with *disneylandization*, and therefore entailing a cultural misuse.

Although within a few years a form of accumulation began to take place (Michel 2020), this did not constitute a solid policy.

1.13.3. *The mayor's decisive commitment*

The idea of a to and fro in decision-making suddenly became evident. The multiplicity of actors and strategies represented a tremendous waste of energy. The director of the tourist office contacted the mayor. From that moment on, Jean-Marc Ayrault would be responsible for decisions. However, he placed Jean Blaise in charge of operations. The tandem worked efficiently; one ignited the flame of ideas, while the other sealed the decisions. On January 1, 2012, they created a new structure with a local public company status. The public company had a budget of 29.5 million euros, including 59% subsidies and 41% revenue. It hired 350 workers, including 295 full-time equivalent (FTE) employees (210 FTE for *Machines de l'île* and the castle). The capital was shared between Nantes metropolis (68%), the city (17%), the region and the department (each community 5%), and the other members had 3% (the city of Saint-Nazaire, the towns of Estuaire and Sillon, Sud Estuaire) or 2% (Clisson, Sèvre and Maine Agglomeration). Jean Blaise insisted on calling it *Le Voyage à Nantes* (the Trip to Nantes), to the great dismay of internal communications specialists. Like the castle or the Machines, the tourist office (in short, every tourism-related actor in Nantes) was invited to integrate the new structure, but this was not so simple: Jean Blaise imposed his conditions. Even though at the beginning, tourism actors were dissatisfied, success quickly kicked in (as shown in Figure 1.12).

The selection was drastic. The Actors' Council – a flexible structure, open to anyone wishing to join in – cemented the bonds. Success bore witness to this. The integration of vineyard actors illustrated the spreading effect sought by Jean Blaise. The *Château de la Frémoire*, headquarters of the Nantes vineyard wine inter-professions' association, became the seat of festive and culinary events, based on the tasting of Muscadet. For Jean Blaise, culture was to transform the city through the deployment of lasting installations. In 2019, 57 permanent works were set up in public spaces, all of which were connected by a route, the green line. Actors wishing to join the *Voyage à Nantes* had to comply with that rule. Therefore, the tourist bus was reinvented, the Okko hotel offered a room *designated* by a contemporary artist, Julien Nédélec (who went so far as to review the standards for setting up a bed), the vineyard hosted the *Porte-vue*, a panoramic terrace set up behind the church of Château-Thébaud, etc. However, resistance persisted. For example, the project to build a crossing at the Liveau paper mill in Gorges came up against regulations and some of the residents who rejected Jean Blaise's ferry bridge scheme. Reflection was pursued in favor of another idea.

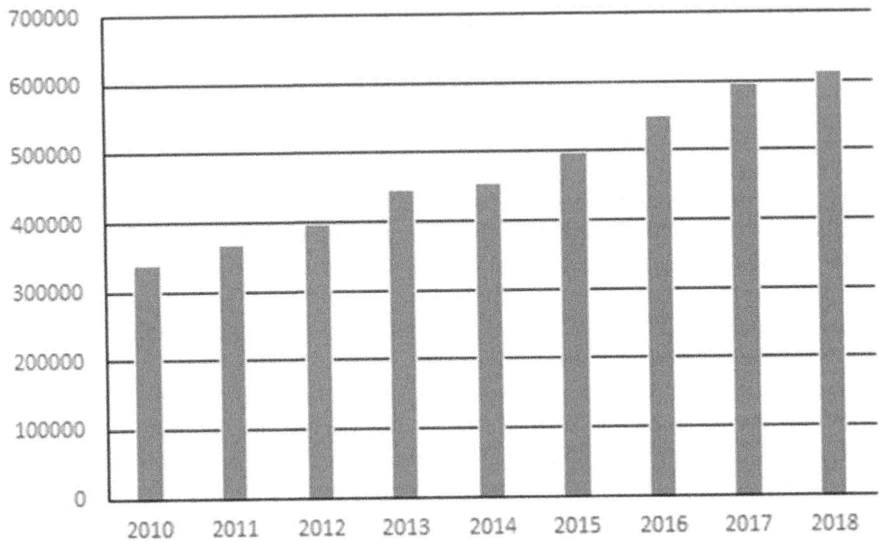

Figure 1.12. *Evolution of the number of commercial overnight stays during the summer seasons*[109] *from 2010 to 2018, according to the tourist tax (source: Voyage à Nantes, creation by the authors)*

1.13.4. *The quality of the places*

The conditions for a successful territorial development through events do not exclusively depend on the actors involved. Behind the good intentions, they are not sufficient by themselves. The quality of places is also at stake. As a matter of fact, it is all of the physical conditions and characteristics of a place that shape its quality and contributes to its construction as a recognized and accessible space. However, the ideal dimension has to be combined with the real. In the case of tourism, this quality is appreciated and sought after by the tourist, who will choose to travel to such or such place in order to spend a moment of recreation in the best possible conditions. In this sense, the place's quality is twofold: it is appreciated not only from the inside, in terms of the visit content offered in the place, but also from the outside, the place's location. Nantes is set in an unfavorable position (a western margin destination, eccentric towards the west).

109 Measurement during the summer rules out the influence of business trips on attendance because business trips present a complementary temporality to that of tourists.

The strategy led by Jean Blaise and Jean-Marc Ayrault comprised two facets. On the one hand, they aimed to strengthen attractiveness through an "intensification" of the place, acting on two levers: revitalizing heritage areas such as the castle (reopened in 2007) and the Museum of Fine Arts (both rehabilitated), and producing new things, such as the *Machines de l'île*, inaugurated in 2007, the *Hangar à Bananes* (HAB, a banana shed transformed from a port wasteland into a contemporary art gallery), or the Memorial for slavery abolition.

On the other hand, they also extended the area by throwing "branches" (according to the term used by Jean Blaise) towards the outside; one towards the estuary, between Nantes and Saint-Nazaire, an in-between space dedicated to wandering among contemporary creations, and incorporating peripheral towns. The other was thrown towards the vineyard, gradually from Vertou to Clisson, which was not obvious at first, between the town on the left and its marked vineyard on the right. The project was included in tourist itineraries, particularly foreigners. That is how the practice of discovery operates through a circuit (Violier 2017), or a network of places along which individuals travel. Only global metropolises, such as Paris, propose stays dedicated to discovery. Following the advice from tourism specialists, then incorporated into the structure, Saint-Malo and Mont-Saint-Michel appeared as milestones in the itinerary "Modern crossing of an old region", spanning the valley of the Loire, passing through Nantes before heading north. Therefore, a metropolis was built with a European dimension, which profoundly modified the place's qualities. In the hotel industry, foreign customers represented 13.4% of attendance, but increased to 19.5% during the summer, which is the tourist season par excellence. According to INSEE, International tourist overnight stays increased by 80% in eight years. Formerly very discreet, Nantes was featured on the tourist map with the *Machines de l'île*, an attraction which welcomed 674,395 people in 2017, placing them at the 42nd rank of "the most visited monuments, cultural and recreational sites", according to the *Tourism Memento* by the GDE (General Directorate for Enterprises, 2018 edition), and the 22nd position outside Paris and its region[110].

With success, also came criticism. This is the case with tourism and tourists: it is always too much or not enough, not enough or too much. In the past, Nantes was not visible, but today the effects of metropolization are attracting criticism, particularly the rise in land prices. However, have we ever seen a dynamic city with falling prices? In short, the city in the countryside. It is up to the actors to imagine the regulations that will allow citizens to thrive in a bustling city.

110 The *Château de Nantes* had a better performance, with 1,471,741 visitors, occupying the 14th position globally and the 4th rank outside the capital. The entry is free of charge, which means that the count is broad, including all passers-by and those returning.

2

Geographic Analyses

2.1. Temporal circumstances

As in any human enterprise, the emergence and affirmation of a tourism system does not escape time. A nuance will be drawn between pre-tourism and the expansion of a Localized Tourism System (LTS).

2.1.1. *Before the founding of the LTS*

The temporal circumstances in which a Localized Tourist System is established will be analyzed in the following section. In certain cases, a critical situation (or an event experienced as such by some of the actors) constitutes the trigger. But this is not a generality.

2.1.1.1. *A crisis at the origin*

In some places, the emergence of an LTS is the result of a crisis affecting the local economy, with a time lag. In Amnéville, it was the decline of steel production, which had economically driven the region since the 18th century, that encouraged the mayor, Jean Kiffer, a doctor by profession, to develop a new activity around a thermal spring. Later on, he enriched the project with a variety of entertainment proposals.

In Nantes, the shift in maritime activities downstream of the estuary was emptying the city of part of its activities. The maritimization process that affected the economy at the end of the 1980s, particularly in France, was quite a blow for the city located at the bottom of the estuary. Culture was immediately seized as an opportunity for new economic impetus:

> It was with the intention of removing this image [engendered by the industrial crisis] and revitalizing the territory that the municipal team elected in 1989 placed culture at the heart of its strategy and the making of the city, conceiving it as an axis for decompartmentalizing urban, economic and social policies. The deployment of this strategy relied on a tight network made up of politicians, developers and cultural actors such as Jean-Marc Ayrault (mayor from 1989 to 2012), Laurent Théry (director of Samoa, a development company from the island of Nantes, from 2004 to 2010), Jean Blaise (creator of *Les Allumées* and *Estuaire*, currently director of "The Trip to Nantes"), Pierre Orefice and François Delarozière (former members of *Royal de luxe* and co-authors of the *Machines de l'île*). The first actions were evident at the turn of the 1990s with the organization of the *Allumées* festival and the reception of *Royal de luxe* company, author of the Saga of the Giants (Michel 2020).

Although the relationship with tourism initially received little attention, it was strengthened after 2011. The local public company *Voyage à Nantes (VAN)* was created with the purpose of bringing together the actors from the two fields, whose shared desire was to promote the city. Jean Blaise became its CEO once the director of the tourist office was dismissed. "The goal is to turn Nantes into an important tourist destination and thus increase its economic growth, by positioning it as a city of art and culture" (Michel 2020).

Until this strategy was implemented – as in most cities in part of Europe, and particularly in France – the set "Old town-Castle-Cathedral" engaged a low-intensity tourist flow (Violier 1998a). In those times, the hustle and bustle that reigned in the neighboring town of Angers, inspired dreams and fantasies[1]. Nantes has now established itself as an urban destination, commensurate with the ambitions of a regional metropolis, a transformation triggered in 2007, with the organization of an event which brings the estuary to life during the summertime. The port's decline provided spaces for the tourism project to flourish, the "Machines" being located along the Dubigeon wasteland on the island of Nantes, which once harbored the old shipyards, closed in 1987.

The critical situation born from the obsolescence of an industrial system is also found in the creation of the LTS run by the Louvre-Lens. The closure of the coal mines spelled doom for a region where ancient buildings and even slag heaps were being declared heritage sites, classified by UNESCO in 2012. Despite all efforts, tourism development did not measure up to the place's quality nor did it meet the standards that the international organization and the region's public authorities were

1 Interview with a tourist office executive discussing this period.

hoping for. This is what happens with labeled sites: while some are highly popular before even getting the distinction by the world organization, others may actually benefit from the distinction, but in certain cases, the expected ripple effect does not come about. The operation for transplanting the emanation of a prestigious cultural establishment attempted to compensate for this.

In other cases, as in Saint-Jean-de-Monts, it is the perception of the situation which reflects the reaction undertaken by the municipality, more than the crisis itself. Even though the station was invented by tourists in the 1850s, the municipality only became aware of this in 1863 and adapted the place to mass tourism[2] in the years 1955 to 1975, by acquiring a beach front with high-rise tower blocks. Its momentum seemed to run out of steam in the mid-1980s. The actors closely monitored decline indicators such as the contraction in the tourist's place of origin and the decrease in average unit expenditure. The city councilors then intervened in two stages. In the mid-1980s, a decision was made to break away from the urban planning model based on verticality and to set up a neighborhood of villas, stimulating the construction of facilities to modify the place's image. To the north of the station, a golf course bordered the homes, whereas a thalassotherapy center was built with the contribution of local capital. At the same time, the collapse of SACOM provided a reason for intervening as well as the spaces required (Vincent 2015). However, the effects of this first stage were considered to be insufficient. A semi-public company was created to support a more global tourism policy.

2.1.1.2. *The crisis ... not always necessary nor sufficient*

In other places, no crisis seems to have occurred and to have presided over change. All seemed to be going well in the world of Cognac, as in the namesake town, when the tourism strategy was engaged. The stimulus came from the outside with the Évin law of January 10, 1991, which banned alcohol advertising. Tourism then became an axis of communication. It is therefore the modification of the environment that was at the origin of change.

In the *Saumurois*, a situation bordering with a major tourist destination is combined with internal difficulties, reflecting the challenges for an LTS to come about. The analysis carried out by Jérôme Piriou (2019) highlights the real toils and troubles for developing a tourist region along the Loire Valley. This was reached by a first displacement made from the place of residence. Its construction was pursued by the secondary mobilities which stretch along the river, and whereby tourists

2 For us the expression is not pejorative and synonymous with disaster or invasion but describes a tourism system characterized by access to tourism for the greatest number of people, which means social progress (Violier 2016). While it is true that the increase in the number of tourists does not only bring positive repercussions, but some solutions do also exist to keep the situation under control.

frequent the different chosen attractions. But the *Saumurois* is integrated into the western margin, and, in fact, it is less frequented than the heart, made up of two sectors (the first located to the east between the castles of Chambord, Chenonceau and Amboise, and the second to the west around Azay-le-Rideau).

Internal dynamics are also at play. In the case study, we detected that the impulse for creating an organized tourist territory came from the outside, more precisely, from the pressure exerted by the Regional Tourism Committee of Pays de la Loire. Instead, local history was rather marked by municipal autonomy, as was evident in the case of the small towns like Saumur, Longué-Jumelles or Doué-la-Fontaine. The dynamics of the latter rely on the "Bioparc", a zoo driven by a dynasty of lively and inspired entrepreneurs, or the *roseraie*, a special rose garden hosting a local festival. In relation to decentralization, the regional institution that was formed in the 1980s exerted pressure within the framework of the plan prepared together with the state, so that a Saumur County was formed bringing together the different entities. The possibility of receiving aid was suspended.

On the contrary, the tourist economy of the main town faced several obstacles. The castle, owned by the town, was not lively and remained poorly integrated into the Loire destination. A differentiation strategy was missing within this concentration of over 50 castles (generally dating from the Renaissance), notwithstanding the fact that the site is exceptional, undoubtedly one of the most spectacular in the region. The former National Riding School, known worldwide under its historic name *Cadre Noir*, plays its own game, based on a show which takes place more often elsewhere than in Saumur itself. As for the wine houses, specializing in the production of delicate bubble bottles elaborated with the traditional method in former quarries, they fiercely compete with one another. Two of them have positioned themselves more clearly in relation to tourism, with a richer proposal than the simple visit-tasting and by setting up dedicated spaces. In its heart, the Bouvet-Ladubay company offers a contemporary art gallery, whereas Ackerman has turned the cellars into both a heritage space (dedicated to the company's history) and a cultural venue, incorporating artistic creations. Around these three cities revolve a large number of places open to visitors, offering accommodation and catering infrastructure. Most of them are small – which does not exclude quality – and welcome a few thousand visitors per year. The Abbey of Fontevraud particularly stands out in this setting. For several years, the abbey has maintained a positive trajectory, opening its doors to the public and strengthening the supply. Furthermore, since it is located at the eastern entrance to the region, it provides a connection with Touraine. Located on the western margins of the famous tourist region of the Loire Valley, what we hesitate to call the *Saumurois* due to the weakness of the polarization by the town, is akin to a vague margin which struggles to concentrate tourists.

Whether the crisis is real, measured, simply perceived or anticipated – but still without any concluding indicators – tourism has been seized as an opportunity to ensure the reproduction of localized society.

2.1.1.3. *Tourism: a lever for local development strategies*

From this perspective of local development, tourism constitutes a particularly relevant avenue (Violier 2008). The activity is characterized by the movement of individuals towards the places, accompanied by a financial flow that goes hand in hand with tourist spending. On the contrary, as tourists stay at the place for a period of time, their spending fuels various activities such as food retailing for example, apart from the expenditure in restaurants. This spread is measured through the multiplier coefficient. The activity is considered as a driving engine, meaning that it leads others.

However, usual statistical techniques do not accurately capture this effect. The tourism satellite account, for example, recommended by the World Tourism Organization for international trade, integrates these characteristic activities and ripple effects as a measurable indicator. But while the former are extensively evaluated – after all, coffee shops and restaurants are not exclusively frequented by tourists – the latter are underestimated. In fact, the list does not include food expenses at retailers or producers within the framework of short tours. However, only a small proportion of tourists take part in an all-inclusive tour or holiday, where the tour operator seems to control most of the tourists' expenditure. Table 2.1 shows that only a minority of individuals choose a tourist trip within the value creation classic pattern that linearly articulates the interventions of travel agencies, tour operators, transporters and accommodation providers. Adherence to this chain, which can be controlled by a single operator in the case of a vertically integrated firm, only constitutes 15% of cases. It is more common when the destination is "abroad or in the overseas departments" (nearly 39%), and much less frequent for "Metropolitan France"[3] (only less than 8%). On the contrary, most bookings directly take place with transport or accommodation providers (in almost 58% of cases). This is largely predominant in the case of "Metropolitan France" and represents the first option for "abroad and the overseas territories" (45.5%), contrasting the sum of the three other possibilities mentioned.

These data certainly need to be nuanced. This national overview is likely to understate the share of package tours in overall tourist expenditure. On the one hand, they present the French case as a territory where individuals are offered the conditions matching conventional social standards (rest with hot baths and sunshine

3 Although this is the terminology used by the source, the qualifier "Metropolitan" is controversial in that it refers to the colonial period ("*France metropole*" was used to convey the difference with the French overseas departments).

guaranteeing a tan). In less advantaged departure countries, and where insularity further complicates travel (such as in the British Isles), the use of all-inclusive organized trips is more frequent. On the other hand, statistics in France generally include visits to relatives and friends under the tourism category. While the cultivation of social bonds can be integrated into the tourists' projects (or even though certain opportunities for cheap accommodation are known), it seems excessive to integrate all of these mobilities under the tourism category. Some of them pertain to social relations which unfold without any bookings nor the mediation of professional actors. Passage through the places is not based on the qualities they present. The experience primarily (or almost exclusively) takes place in the hosts' homes. Hence, we prefer to use the term "travelers" rather than "tourists", which should be subject to a more restricted use (as in the concept developed within the MIT team, Knafou and Stock 2013).

	Metropolitan (mainland) France	**Abroad or overseas territories**	**Total**
Trips booked (in millions)	69.2	22.3	91.5
Travel agencies or tour operators (in %)	7.7	38.3	15.1
Service providers (in %)	61.3	45.5	57.5
Others[4] (in %)	36.3	44.8	38.5

Table 2.1. *How do travelers organize their trips? The total for each destination is greater than 100%, because the modes can be combined*[5] *(source: simplified table based on* Key tourism figures, *General Directorate for Enterprises, 2018 edition, data from 2017)*

The analysis of the real beneficiaries of the activity is debatable, particularly on an international scale, a scene in which accusations of neocolonialism are frequent. While it is not the purpose of this work to make a state-of-the-art on that issue, the question is also raised locally, and the unequal distribution of profits is an argument opposed to tourist projects. However, indigenous people perceive economic gains through the creation of jobs, whether within local structures or the subsidiaries of large groups. They also have access to opportunities in catering, guiding, the production of objects acquired as souvenirs, in the production of short-term rental accommodation, etc. During the Covid-19 pandemic, the press could truthfully relay

4 Booking platforms, offer comparators, search engines, etc.

5 A person can book accommodation directly with the provider and transportation using another method, for example, through a platform.

the distress of local actors deprived of the tourist windfall after the interruption of arrivals. This proves that at least partly, indigenous artisans are grafted to the actors of tourism systems.

2.1.2. *The itinerary of places*

A place more or less follows the course of time. This is not linear or smooth, but experiences accelerations or ruptures at certain moments, to which the place and society do not adapt (or do so with difficulty). Our starting point will be a criticism of the model proposed by Richard Butler (1980), and we will suggest a different time reading of tourism areas.

2.1.2.1. *Criticism of Butler's model*

Contrary to the assertion by Richard Butler (1980) and his apologists, pace is not determined by the evolution in the number of tourists. In his model, the dynamic is related to the carrying capacity, which – due to saturation – reaches a critical point at a certain moment. This leads to several possible bifurcations. Regardless of this, the author's proposal has been reduced by analysts inspired in his concept of "decline", which was not exactly the author's point. Actually, Butler contemplates various possible itineraries, including rejuvenation (Butler 1980). But let us retain the essential: according to these researchers, the crisis is derived from the confrontation with too many people.

This model can be criticized in several ways. First of all, the tourism area is reduced to a single entity. Until the end of the 1990s, this posture was common. In his work devoted to tourist stations (1996), Vincent Vlès included Paris as well as La Baule and Lourdes under that category. However, this does not correspond to the historical reality of tourismification. Similarly, Daniel Clary (1993) introduced the "station system" as the archetype of the tourist place. As shown in 2002 (MIT team 2002; Duhamel 2003, 2018), a typological approach draws a difference between several categories of places. Stations and resorts (created by and for tourism) are opposed to historically constituted cities like Paris, or isolated sites like Chambord, which are turned into a tourist attraction at some point in their itinerary. This diversity carries the seeds of differentiated relationships between the actors among themselves and with tourism as an object. In the places created, tourism constitutes the essential, and sometimes even the sole economic driver. Therefore, the attention paid by legitimate actors is considerable and technicians benefit from strong support. On the contrary, in a multifunctional place, competition is tough and elected officials can call into question the sector's legitimacy.

Even admitting that the destination is subject to contingencies, these come as much from the outside as from the inside, and do not depend on the carrying capacity (number of individuals welcomed). On the contrary, tourists are lacking when changes that take place within the society lead to broader changes, to which tourism areas are compelled to adapt. However, the nature of this evolution jeopardizes them to a lesser or greater extent, as the capacity to respond to changes is unequally distributed between the actors systems in the places.

Over time, tourism areas accumulate a spatial capital from which relay trajectories (the area still draws tourists in) or metamorphosis (conversion into another dynamic) can occur. Even though the one leading to the abyss cannot be excluded, it is not certain either and, in any case, the cause cannot be entirely attributed to the exceeding number of tourists (Darbellay et al. 2011).

What are these temporalities?

2.1.2.2. *Mutations born at the heart of tourism*

Developments are of different natures. Some take place at the heart of tourism. Others relate to mutations in the values held by society.

2.1.2.2.1. Mutations within tourism

Tourist practices evolve within society. For the coastlines, the most remarkable change was the passage (first, at the beginning of the 20th century in the United States and then in Europe), from the therapeutic cold bath and white skin aesthetics to the triptych consisting of the hot bath, tanning and pleasure (MIT team 2005, 2011; Duhamel and Violier 2009). In the mountains (Deprest 1997), the advent of skiing was also a major issue, because it brought about a change of season, from summer to winter, together with the need to equip accommodation with heating. This posed significant technical challenges for climbing the slopes. However, resistance becomes manifest everywhere, and adaptation requires tackling more or less formidable challenges. In some cases, the end of tourism has been foretold, but this could be due to the disappearance of tourists altogether, not their excessive numbers. Attendance towards destinations like the Côte d'Azur hasn't ceased to grow, and its transformations have been remarkable, despite Butler (1980) considering that the "denaturation" of tourism areas is combined with "overcrowding", and therefore, is at the origin of decline. But why should we assume that a fishing village is more natural than a tourist station? In both cases, these are human constructions, produced at different moments and responding to specific functions. The challenges are more or less daunting and the LTS more or less able to meet them.

The advent of the "Destination Anjou" LTS also reflects a mutation in tourism. Its creation was motivated by the competition exerted by large hotel chains against independent and family establishments. This development is derived from the rationalization of the supply and a classic concentration process, signaling the passage from an entrepreneurial activity at the artisanal stage (marked by the autonomy of small establishments), to an industrial stage (characterized by the emergence of networks in the field of tourism). Unlike other economic sectors, concentration does not take place through in situ accumulation, but by the creation of canvases capable of welcoming travelers wherever they travel. At the local level, the sluggishness of the Hotelier Club was an adjuvant. The strategy then was to boost marketing through a heterogeneous network not only offering accommodation (chains included, because their establishments cooperate at the local level), but also objects with complementary leisure activities and restaurants, within the perimeter of the Department of Maine-et-Loire, immersed in the logic of an institutional destination.

2.1.2.2.2. System mutations

But disturbances came from the system as a whole, and not only from the evolution of one component, as we have seen. In other words, tourism is a meta-system which has evolved over time (Violier 2016). It began by only affecting a minority of wealthy and idle individuals who could devote their time to organizing traveling, because there were no dedicated organizations capable of doing so. Furthermore, they reached to their family or network of friends for accommodation. They were a few in number and relied on the narratives of their predecessors for guidance. From 1830 to 1850, the system changed and passed to the small-number stage. As Laurent Tissot (2000) has shown: the printed tourist guide was invented, the tour operator entered the scene, the railway was diverted from its industrial use and, finally, the tourist station was invented. Places shifted from the building of a few villages every year to housing developments. After the 1950s in Europe, enrichment and redistribution favored the emergence of mass tourism[6] (understood as the stage reached by a society within which the practice becomes widespread until it is widely shared, even though some remain excluded)[7]. Since the 1980s, the Western world has shifted to personalized mass tourism, a stage marked by a greater

6 We use the expression in a scientific sense, characterizing a socio-technical system deployed in a society having reached such a level of economic development that tourism becomes accessible to the greatest number, that is, to at least 50% of the individuals who compose it (Violier 2016). On a more political level, this social dissemination seems desirable to us. On the contrary, a pejorative use is made by the authors wishing to point out a destructive effect which has never been demonstrated. Although environmental effects are not negligible, we must acknowledge the fact that no human activity escapes these issues and that tighter regulations can be sought.

7 Not to mention that tourism is not an obligation and that individuals may give it up temporarily or be permanently homebound. The share of excluded people can be estimated in France at 15% of the residents.

diversification in the ways of being a tourist and the discourses evoking the experiences sought or lived.

These different stages pose various adaptation problems in tourism areas. In some of them, the ease of access means that success is immediate and therefore, difficult to manage. In other places, fairly simple means can be engaged. For example, Mauritius manages attendance by adjusting the frequency of the air connections. The Saint-Jean-de-Monts station reached the exhaustion of its mass tourism phase (for which it was configured between 1955 and 1975) and must now transition to a diversification of demand.

2.1.2.3. *Societal times*

Developments that also weaken tourist destinations can emerge within society.

2.1.2.3.1. Changes in the urban planning model

The production of accommodation in the years 1950–1970 was based on an urban planning model (the adaptation to automobile traffic) and an architectural one (functionalism and collective high-rise tower blocks). Although it was designed in large cities, it became widespread in tourism areas. This development corresponds to the access of European society to mass tourism and the challenge of tourism's carrying capacity. For the most part, the characteristic villas of the *Belle Époque* were destroyed, except in the places in decline, where they remained as witnesses of a bygone era[8], or in spaces where an outdated bourgeoisie has taken refuge (as in Dinard). In other places, relics survived through the effects of succession and family dynamics. They are now integrated into the vast movement of heritage extension (references), a development that contributes to the dynamics of tourist places, creating the conditions for discovery in a tourist universe dedicated to rest. For example, La Baule-Escoublac[9] was classified as an urban and landscape architectural heritage protection zone in 2006, and then transformed into an Architecture and Heritage Enhancement Area (AVAP, *Aire de mise en valeur de l'architecture et du patrimoine*) on February 22, 2013. While this policy sought to protect around 6,000 villas, it was implemented only after a phase of destruction affected the seafront (where the adaptation to the times of mass tourism concentrated and was translated by the construction of six- to eight-story collective buildings).

---

8 The frequently used term "preserved" is downright inadequate because preservation requires a strategy, whereas in reality, these sites constitute abandoned residues.

9 Official toponym of the municipality which was invented in 1961 by combining the station's name, La Baule (derived from the old local term "bôle"), with the name of the former municipality of Escoublac.

2.1.2.3.2. The advent of sustainable development

Other developments are derived from changes taking place in terms of social values. After the 1970s, sustainable development gradually took precedence over the Promethean myth of limitless transformations of the planet (Pickel Chevalier 2014). In that case as well, places were requested to conform to the same more or less effective adaptation mechanisms, depending on the nature of changes and the LTS' capacity to integrate the new expectations.

Thus, the seafront boulevard in Saint-Jean-de-Monts, established in the 1950s, was completely redesigned. The car space was reduced to two lanes and the car parks were moved in order to free up space for separate pedestrian and cycle tours, with a sea view, whereas a track network made the forest accessible to bicycles. The buildings were not destroyed, since there were numerous restrictions to do so. In addition, straightforward action required the agreement of a large number of owners, not all of whom wished to invest. But the new extensions were made taking into account the new provisions introduced by the coastal law, passed in 1986, away from the coast and integrating the new standards, namely individual housing. In the municipality of Saint-Hilaire-de-Riez, the extension of the seafront towards the south – already underway – is fixed. The two sets already completed, "les Becs" and "les Mouettes" are fixed in the middle of the dune massif.

2.1.2.3.3. Institutional changes

Other developments also affect the institutional domain. The territorial decentralization implemented in France since the 1982 reforms is characterized by a transfer of power to sub-state entities. In terms of tourism, this was marked by the capacity of local institutions to carry out large-scale projects. As from that moment, mayors played a significant role. Apart from Paris – a city visited by the English from 1820 after the return to peace, until the 1970s – cities in France did not undertake vigorous action in terms of tourism policies. Even though the practice of discovery existed, this was limited and concentrated around a few monuments, cathedrals, castles and museums, or groups (in other words, the historic districts rehabilitated within the framework of the Malraux Law of August 4, 1962). The affirmation of entrepreneurial mayors focusing on local development promoted several cities, notably Lille, Nantes, Marseille, etc. (Morice and Violier 2009; Violier et al. 2021). However, avant-garde initiatives were also carried out, such as the Angoulême International Comics Festival.

With the NOTRe law, passed in 2015, tourism became a municipal competence. Until that date, the possibility existed, but had to result from a favorable vote from the municipal councils. The new provision was evidently favorable to LTS creation, as it provided a regulatory framework for the emergence of a larger, non-divided perimeter, better adapted to the reality of tourist practices. However, this implementation resulted in different geometries and the adequacy between a tourist destination, as we understand it[10], and a municipal territory has not systematically been a priority, even in cases where tourism is the main economic driver. In the various cases studied, distortions appeared throughout the development of projects. In certain situations, the municipal marquetry played a role.

In Amnéville, the question was solved by the voluntarism of the mayor, who did not hesitate to integrate territory portions from neighboring municipalities without further ado, an initiative which sparked tensions. As for the Pays de Monts, in the north of Vendée, the community of municipalities was based on the cantonal perimeter. This approach made it possible to unify the management and dynamics of tourism areas sharing a common history. Even though Saint-Jean became involved in tourism earlier (as from 1867), this was done quite modestly and under the municipality's initiative, although bathers had frequented the place since 1850. Diffusion then took place both towards the north (in the territories of Notre-Dame-de-Monts and La Barre-de-Monts) and the south (Saint-Hilaire-de-Riez). In the past, a coastal railway served the five municipalities. Expansion resulted from the transition to mass tourism, something the elected officials chose to support locally in a very proactive manner after the 1950s. This trend was later followed by neighboring municipalities. The same company, SACOM, brought them together (Vincent 2015). But cantonal boundaries do not correspond to tourist practices. A Pays de Monts destination should have included Saint-Hilaire, which shares the beach with Saint-Jean-de-Monts and, thanks to the *corniche*, offers complementary sightseeing and discovery possibilities. In addition, Saint-Gilles-Croix-de-Vie provides the maritime component. The perimeter of the ephemeral "Vendée des Iles" county was highly relevant with the islands of Yeu and Noirmoutier (destinations for excursions), and even the municipality of Sallertaine, integrated into the canton of Challans, where many activities attract tourists from the surrounding area, including canoe trips in the Breton Marshes. However, the institutional system here fragmented the destination.

Whenever the department imposed itself, either through the departmental council or the chamber of commerce, no coherence in terms of tourism was ever sought. Nevertheless, as will be discussed further, this framework seems too vast to host an LTS. A formula making room for coordination would be preferable. In the case of Cognac, institutional changes undoubtedly slowed the emergence of an operational LTS. First of all, in 2013, the Regional Tourism Committee in the Region of Poitou-Charentes was abolished, effectively reducing the resources made available to an embryo which could have led to the founding of a more effective LTS. Regional redistribution began in 2015, integrating the four departments into

10 The discrepancies between tourist spaces constructed by tourists and institutional spaces will be discussed further.

the very vast entity of Nouvelle-Aquitaine. Until the end of 2017, energies focused on the implementation of the new institutional framework. The subservience of tourism policies to the administrative framework led to harmful interference for destinations.

The transfer of the tourism competence to the municipal level does not pose a problem, but actually depends on the local situation. In Saint-Jean-de-Monts, it took two years for the name of the main station to be recognized as the best standard of promotion by the elected officials from other municipalities. But this is not always simple, and a recognized destination runs the risk of finding itself trapped within a group where tourism is not the main economic driver.

A recent legislative development confirmed that this posed a problem and relaxed the provision. Article 10 of French Law No. 2022-217 from February 21, 2022 stipulates that "one or more municipalities classified as tourist stations [...] may claim to regain the exercise of the competence over "tourism promotion, including the creation of tourist offices". The restitution of this competence shall be decided by concordant deliberations from the deliberative body of the urban community and the municipal councils of all its member municipalities"[11]. It is not certain whether this will be enough to ease tensions, as a single vote can prevent a return to autonomy.

In short, the sustainability of tourist areas is jeopardized by historical, global and tourism developments, but the spatial dimension also introduces certain issues and questions.

2.2. The circumstances of places and spaces

Even embracing the fundamental concepts of geographical science, our goal is to analyze how the tourism approach calls them into question and then to reflect on the implications this has for our work.

2.2.1. *From the tourist place to the tourist destination*

The notion of tourist destination deserves to be questioned. Often invoked in writings (sometimes in an outrageous manner), it is also used on an institutional level to refer to the organizations responsible for tourism. Is this notion interesting? Can we turn it into a concept and use it to help us think about Localized Tourism Systems as a complement to the place, a key concept in geography?

11 This designates the tourist station in the regulatory sense of the term, and not the concept developed within the framework of the geographical approach to tourism (MIT team 2002).

2.2.1.1. *The tourist place*

Place and space are two fundamental concepts in the discipline. By definition, and by excessively reducing a fruitful debate, the concept of space is opposed to that of place. While the first one is characterized by extent, the other features the absence of distance (for further information, refer to Lussault (2013)). A place or space is touristy when it is transformed by tourism. The fact of traveling, by itself, does not suffice. In countries which have been extensively visited and for long, it is possible to meet tourists everywhere, mainly because the connections to the localities included in the program require twists and turns.

	Situation before tourism	Economic system	Tourist function(s)	Examples
Site	Previous creation	Specializing in tourism	Specializing in leisure or accommodation	Castles of Chambord and Chenonceau
Tourist city/town	Previous creation	Diversified system	Polyvalent	Amboise
Touristified village	Previous creation	Became specialized in tourism	Specializing in leisure or accommodation	Sallertaine
Stopover city/town	Previous creation	Diversified system	Specializing in accommodation	Towers
Station	Created by tourism	Specializing in tourism	Polyvalent	Saint-Jean-de-Monts
Resort	Created by tourism	Specializing in tourism	Polyvalent	Center Parcs

Table 2.2. *Typology of tourist places applied to the case studies from Chapter 2, according to the typology of tourist places conceived by the MIT team (2002) (Duhamel 2018)*

After making this point clear, places were analyzed by geographers. For his part, Vincent Vlès (1996) did not make a distinction and generally considers them as stations: places where tourists stop by. In 2002, the members of the MIT team proposed a taxonomy drawing a distinction between the places created by tourism and the ones historically constituted and invested as such (Table 2.2). From the perspective of this work, this duality takes on a complementary meaning. In one case, at least initially, tourism is the only economic driver of the place. The place of the activity justifies a seizure of power by its key players. Furthermore, when the graft is implanted on a clearly different topographical site, the evolution can even lead to separatism from the original village. This is particularly the case for seaside areas far from towns and agricultural villages located inland. For example, Le Touquet, a station on the Picardy coast, was established as a municipality and separated from the original municipality of Cucq in 1912. In other cases, an inversion of centrality occurs to the point of moving the main town hall and even changing the toponym. In La Baule-Escoublac, although the old village's toponym was preserved, it was placed in the second position. Diversification took place with the arrival of new residents, particularly retirees. This is really a game changer. Tourism is no longer the sole concern. In places which have been historically established and invested as such (e.g. cities), tourism must deal with other economic issues. The actor games that we will study in section 2.3 become more complex. Different strategies can compete, among which welcoming temporary residents does not necessarily prevail.

2.2.1.2. *The tourist destinations*

Still, while this typology is relevant for an objectified approach to places, it is not operational for the analysis proposed in this work. Regardless of the places themselves, what we aim to understand is the games of the actors which develop in that context. It is for this reason that we will use the term destination.

2.2.1.2.1. For a scientific concept

The notion of destination has not been explored in depth but used as a blanket term. Used numerous times in certain articles, its meaning has not been explored in depth and would seem to lack interest. Let us provide some preliminary elements. A first and thorough approach was proposed by Hazebroucq (2009) and then Violier (2009) in the same work. They suggested the idea of a double temporality between the tourists' project and its implementation. In doing so, they differentiated between what amounts to representations and to the confrontation with reality. Botti et al. (2019) also share this view. But on re-elaborating the said proposition, they invoke the typology of tourist places and include the destination under the narrow dimension of places, which are, in the sense of geography, elementary units. This position is partly understandable since their project is to address the question within the framework of management sciences, whose object is the study of organizations and institutions, which are construed over territorial bases. But it hinders understanding in their full complexity the relationships between the different tourist organizations created at all of the levels of the system (sometimes in competition, sometimes as a complement).

In point of fact, the project – the cognitive dimension – is not necessarily limited to one place, or even to several successively inhabited areas. For example, a route along the Loire Valley involves various places visited one after another, which may have different, but complementary, functions (Piriou 2019). Several sites are solely dedicated to discovery (such as Chambord or Chenonceau), others are only stopovers, where individuals spend a night or two (in Tours, for instance), or cities may combine accommodation and the visit (such as Blois and Amboise). In each of these places, the expression used is "transit tourism", which does not correspond to the reality observed at the scale of the tourist region (Piriou 2019).

On the contrary, implementation as well as the confrontation with reality and otherness also translate the production of a lived space which differs from initial representations. Even though anchoring is more frequently combined with the practice of rest (Violier 2017), this does not necessarily synthesize the whole experience (e.g. an establishment within a Center Parcs). Such an experience may open to other places, in the form of excursions and for visiting purposes only. Jérôme Piriou (2012, 2019) showed that an establishment along the Emerald Coast does not prevent an escape to Brest and, why not, a visit to its Océanopolis theme park. In a survey carried out in 2012 in Saint-Jean-de-Monts, the on-site stay was supplemented by wanderings not only towards the islands of Yeu and Noirmoutier but also towards Puy du Fou, located an hour and a half away by road (at that moment, the Saint-Jean-de-Monts tourist office was the first sales point for entry tickets, after the establishment itself[12]), La Rochelle (an hour and a half away by road) and Nantes (one hour away). The initial project, based on rest, can be satisfied more quickly than imagined. From then on, individuals try to occupy themselves differently, and, as a consequence, start moving around. This reality disqualifies the overly simplistic opposition suggested by Jean-Didier Urbain between "vacationers" and "real tourists"[13].

While the individuals' project tends to be dominated by discovery, secondary mobilities[14] can be multiple. This practice is more often associated with a circuit or loop (Violier 2017), which is a network of several places visited successively. However, a stay for this practice cannot be ruled out, as it is frequent in the discovery of metropolises featuring a concentration of objects to view, for which several nights on site are needed. All the same, excursions to sites can be organized as part of a rest stay (a visit to the Great Wall from Beijing, to Ellis Island from New York or to isolated venues like the Palace of Versailles or Monet's gardens).

The question of the destination has been addressed by Kadri et al. (2011):

12 Unpublished investigation report for the Saint-Jean-de-Monts mixed economy company.

13 "*1936–2006 : il était une fois les vacances*", interview with the anthropologist Jean-Didier Urbain, Le Monde 2, June 10, 2006, Pascale Krémer.

14 As different from "original mobility", which designates the route taken from the residence (or the previously frequented space, which may have been visited with a tourist intention in mind), towards a destination.

> However, and despite the effort to overcome one-dimensional and reductive readings of the concept of destination, the project (for the authors quoted, Hazebroucq and Violier) only identifies one component of the overall destination. We think that it is necessary to talk about the destination in terms of "projects", in the plural and not in the singular as the above-mentioned authors do. Our hypothesis is that the destination presents itself as a dynamic set of projects, conceived as intentional and interested efforts by the actors involved in the construction of the destination. These projects, simultaneously differentiated, but contributing to the formation of an integrated project, comprise various dimensions. Hence, we would like to distinguish five projects organically related to one another: an anthropological project (the vision of *leaders*, the motivation of tourists; an economic project (market, tourist product); a development project (tourismification of spaces); a management project (organization and governance mechanism); an urban project (societal project).

More recently, a reference to this approach has been made in the *Vocabulaire du discours touristique* (Kadri et al. 2022, pp. 108–113). There are, however, two limitations to this approach. It is both too broad (because it confuses several spaces, notably those perceived and produced by economic or institutional actors) and too narrow, in the sense that it favors the territory of institutions ("urban project", "development project"). On the contrary, the heuristic value of the destination concept lies in its focus on individual tourists. This distinctive approach makes it possible to highlight several strategies. On the one hand, professionals try to capture the attention of tourists, whose mobility potentially poses multiple problems. On the other hand, legitimate actors, and their technical operators at different levels (among whom there are divergences), attempt to carry out territorial impositions. On the other hand, tourists are rarely present in this approach. They merely appear within the "anthropological project" through "their motivations". Instead, tourists have intentions constructed in accordance with their representations and, once territorialized, implement them in what they consider to be the destination.

The destination is therefore a place or space with two dimensions. A cognitive one, constructed from the representations of individuals. The images provided by the media, and primarily by their loved ones (and now by social networks and influencers), participate in their construction. They are possibly informed by previous trips. The other dimension is drawn from experience and the project's implementation. In this sense, the destination can be more than a place. It integrates all of these elementary units inspired in tourist representations and converge into a whole apprehended during a tourist trip. In this way, the destination has dimensions which vary from tourist to tourist, who are after all, socialized individuals.

Box 2.1. *Destination*

2.2.1.2.2. A question of size

However, the scope of the destination is variable, particularly in a tourism approach. Since the concept of place pragmatically translates a relative relationship, in space the destination is constructed by tourist practices and by the interventions of local actors (even though they are pieces belonging to national and international networks which try to influence the choices made by the former). Is there an ideal dimension from the point of view of the construction of Localized Tourism Systems? This question can also be formulated as follows: what does geographical proximity entail in concrete terms?

The lower dimension, the one below which geographical proximity is not relevant, is measured according to the emergence of a diversity capable of guaranteeing functional complementarity suiting tourist needs. Deterritorialization requires access to accommodation, food supply and suggestions that can be used to occupy free time and help in accomplishing individual projects, while leaving room for the unexpected. Indeed, tourists are characterized by variable but extensive availability. By definition, they only have leisure activities, the ones that the dominant practice is supposed to fill. But most are jacks of all trades, and the main practice governing the choice of destination (in terms of time spent or the priorities established by the individuals concerned), is far from exclusive. On the contrary, extremely varied combinations are formed (Mondou and Violier 2009; MIT team, 2011).

They are also permeable to the suggestions of the place. The latter can be latent proposals (already there, but ignored) or new things (created or having emerged spontaneously, sometimes in a completely uncontrolled manner). Thus, the grounding of a cargo ship, the *Artemis*, on March 10, 2008 on *Sables-d'Olonne* beach suddenly enhanced the leisure activities of individuals. The date was not optimal from the point of view of the season, but for communication, it was rather timely, as evidenced by the following article excerpt:

> In the meantime, the cargo ship delights the many, many onlookers, who come to see the helplessness of men in the face of the omnipotence of nature. The town of Les Sables seized the opportunity to communicate this unusual event. […] The town and tourist office websites had a burst of activity. Over 100,000 connections were estimated on the two sites combined, at a time when the Internet was not what it is today[15].

15 Article written by Brendan Martineau (2021). Available at: https://actu.fr/pays-de-la-loire/les-sables-d-olonne_85194/il-y-a-13-ans-the-artemis-was-stranded-on-the-grande-plage-des-sables-d-olonne_39989783.html (Accessed on April 27, 2021).

By definition, tourists have a lot of free time. The principle of club-type accommodation is to offer entertainment to individuals who, suddenly freed from a constrained schedule, find themselves in a position to manage this spare time. The minimum dimension of proximity lies in the capacity to offer diversity and complementarity capable of meeting tourists' expectations, in function of the accomplishment of their project and its day-to-day adjustments. The Center Parcs proposal by Pierre&Vacances intends to meet every need a customer may have. They offer accommodation in bungalows. Catering is available in the form of restaurants within the bubble, but meals can also be prepared by the guests at a small self-service station located in the same place. Additional services in the form of rental equipment suggest the organization of raclettes, fondues, etc. In addition to bike rides (with rental equipment, if needed), leisure activities are shared between aquatic pleasures and a range of services, either at Aqua Mundo or outside the premises. The use of external service providers completes the range, depending on the existing skills in the environment and opportunities. Thus, in Hattigny (a municipality where a Center Parcs is located in Moselle), the local market is the fruit of an initiative by local farmers, who partnered up with the establishment. In this case, the place is a resort (Table 2.2), but remains open to the exterior, following the local management strategy, autonomous on this point. It is therefore a polarized mini network. On the contrary, catering has been entrusted to a service provider (the Areas France company), holder of the contract for all Center Parcs in France.

2.2.2. *The ideal size*

Is there an ideal size on a small scale? Case studies help to answer this question. In fact, the intensity of exchanges within a Localized Tourist System suggests that the relevant crucible should comprise around twenty municipalities, with a clear sub-departmental dimension. Working together presupposes mutual knowledge, which exists or is constructed, a frequency for meetings, and this in combination with traveling times. Whether part of a chain or an independent company, most tourist establishments are characterized by a small size and a limited team, and where strategic decisions are based on a manager already in high demand. The effective participation of actors providing services requires a moderate distance.

Thus, the initiative from the Department of Morbihan was doomed to failure because its dimension was too vast and omitted the strategic choice of integrating the territory's interior (which, insofar as it falls under the scope of the institution, is not likely to engage coastal stakeholders, who have other things to worry about). On the contrary, one or more LTSs, bringing together municipalities and actors from the more or less distant hinterland, could have reflected upon the complementarity of taking advantage of the tourists' summer presence along the coast, rather than presenting this as an unbearable inequality, attributable to tourist blindness. What is more, the systems of actors were operational, but within a restricted spatial dimension, since they relied on mutual knowledge, and above all, did not necessarily respect the departmental boundaries. However, the notion of boundary does not have the same representation for an elected official as for a citizen, and even less, for a tourist. LTSs already at work were simply ignored or compelled to fall into line.

Regardless of "distance" stricto sensu, the appropriate dimensions, areolar or linear, depend on the projects and practices of tourists, as well as on the supply. The practice of discovery along the Loire Valley is organized as a linear tour because tourists come to visit the castles, which are located near the river and the confluences and follow an order inherited from the choices that governed their edification. Along the coasts another form of linearity is imposed by the search for proximity to the beaches and interrupted by marshy or rocky areas, which can give rise to other proposals in favor of tourists. Along the Vendée *corniche*, for example, rest gives way to discovery through walks and contemplation, whereas in the Breton Marshes, in Sallertaine, canoeing is proposed by the *Route du sel* company, combining discovery and body movement. Tourists staying at Saint-Jean-de-Monts or Notre-Dame-de-Monts stations to rest can enrich the combination of their practices thanks to the supply available in this neighboring municipality. These changes in practices result in discontinuities in accommodation concentrations, which are polarized by the stations facing the sea. But they come up against the effects produced by the boundaries between territories. Communication and tourist signage are not always optimal for indicating and facilitating access to these possibilities.

A destination can be summarized into a single geographical location, and also extend towards a network and include different types of places. Part of these are areolar and structured around a hub where accommodation is set. Others are linear. There are also combinations between the two ideal types. It is the mobility of tourists, in relative proximity, which delimits its dimensions. But it is also each individual or group, family or friend – whose members, at a given moment, share an experience for a specific time interval – who outline the lived reticular space. Hence, the tourist destination is the superposition of all the itineraries followed, which combine several types of places (Table 2.2), as it becomes evident in the destinations analyzed in the case studies (Chapter 1).

For example, the tourist region (Piriou 2019b) of the Loire Valley, primarily exploits the image of castle heritage and combines isolated sites (castles like Chambord and Chenonceau and other monuments like the Abbey of Fontevraud), tourist towns (Blois, Amboise), with stopover towns, such as Tours. The latter is rarely visited, in the sense that the establishments located there have a low attendance compared to the emblematic flagships (Figure 2.1). Tourists profit from accommodation there, which generates a bit of evening entertainment downtown, where the restaurants are concentrated. There are also different, even non-heritage, proposals, which can be useful for preparing a diversification strategy. Indeed, the Stendhal syndrome awaits tourists, who are confronted with an accumulation of castles. Despite all being different, they share certain regularities (such as the exterior appearance, combining Tuffeau stone and slate), which can eventually become boring. The challenge of other proposals is to succeed in encouraging tourists to extend the length of their stays. Diversification can be developed within the castle itself, as in Cheverny, which promotes the work of the designer Hergé, who was inspired by its facade to design Moulinsart (especially its exterior). In the same spirit, the Beauval zoo (located in Saint-Aignan, less than 50 km to the south), is easily accessible from the most visited castles.

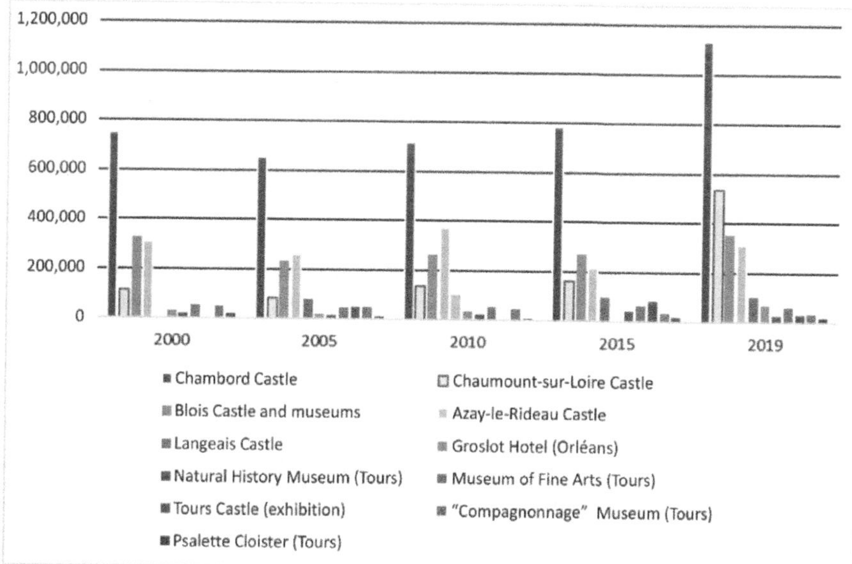

Figure 2.1. *Comparative evolution of attendance at tourist and leisure establishments in the Loire Valley from 2000 to 2019 (source: Regional Tourism Committee of the Region of Centre-Val de Loire). For a color version of this figure, see www.iste.co.uk/louart/touristdestinations.zip*

But at the same time, diversification complicates the development of a common strategy. The actors are torn between the creation of a network of similar entities which will have more strength (e.g. for promotional purposes) and a grouping of heterogeneous entities (more complex to manage but guaranteeing a better response to tourist practices). All things considered though, the majority display a jack-of-all-trades behavior and fail to focus on a single object or even a single practice (MIT team 2011).

2.2.3. *The tourist place in all its forms*

Beyond the boundaries and the perimeter, the question arises as to the state of the place that will be conditioned for a tourist experience. This question has tormented geographers since they scientifically addressed[16] the theme of tourism back in the 1960s (Lazzarotti 2010). The responses varied over time. And different approaches, disqualified by recent research, continue to be invoked nowadays, despite the demonstrations and analyses which have attempted to clarify this question.

2.2.3.1. *Putting an end to outdated ideas*

The first works focused on the latent state of the place which only asked to be revealed, according to processes which have never been explained. Two terms were featured. On the one hand, places seemed to be endowed with a certain "vocation". And this, in function of an "already there". On the other hand, the "potential" delimited the alleged "attractions", which were divided into two categories: the natural and the cultural.

These approaches have been refuted. First, the dichotomy is meaningless, because while the given elements (not constructed by man) obviously exist, they are selected by the view of individuals according to social values. Thus, the Uluru site in the center of the Australian bush is considered as a monolith by Europeans and has been the subject of pedestrian climbs for a long time. On the contrary, for the Indigenous Australians, it is a sacred object. They consider trampling on it an offense and obtained a ban for this in 2019. Even within Western society, the most popular beaches were initially those of the Channel coasts, suitable for therapeutic cold baths, until the beginning of the 20th century, when warmer seas became fashionable. Thus, the same element was evaluated differently at successive times. The notion of "attraction" denies individuals their characteristics as actors, since it is the elements that exert attraction towards humans.

Second, as we have defined it, a destination is a space within which tourists plan to implement a project and where they carry out practices that contribute to recreation. The work of historians has enabled certain geographers to elaborate another tourismification process. The first stage involves an invention, in the sense of a new perspective taken by pioneers. As a result, it provokes an influx of individuals who seek to settle in places which have not been designed for this. Therefore, they accommodate themselves as best they can. In towns like Dieppe, located at the end of roads leaving the towns, they stay at existing inns (*auberges*) for the needs of travelers. Gustave Flaubert and Maxime du Camp have railed against the deplorable state of these improvised havens. Elsewhere, in Arcachon as

16 The first geographer to have flirted with tourism was seemingly Maurice Le Lannou, and he did so by writing a tourist guide.

in the Pays Bigouden, they relied on the hospitality of local inhabitants, fishermen, resin workers or customs officers, who bumped into a commercial venture as soon as they realized that the cycle had repeated itself summer after summer. The change of perspective was at the heart of this mutation. The demonstration was carried out in 1992 by Rémy Knafou, following the hint suggested by Philippe Joutard in 1986.

In a second stage, the now regular presence of tourists gives rise to investment in accommodation infrastructure, initially modest in an artisanal mode. Then, as the phenomenon intensifies, it induces a transition to the industrial stage. Historically, this dynamic has been reinforced by technological progress. Horse-drawn transport gave way to railways in the middle of the 19th century, which considerably increased the number of individuals on the move.

A tourist destination is therefore one or more networked places, represented by individuals as likely to enable them to attain the *recreation* sought.

Box 2.2. *Concept*

The development of communication or marketing techniques, like adaptation, participates in the amplification of the process. But its origin is in a condition that corresponds to the representations of the moment, bearing in mind that by moment we mean a long period of time (one to several centuries) (MIT team 2005). In fact, the cold bath dominated the relationship with the sea for two centuries, whereas the hot bath has been trendy since the beginning of the 20th century. Likewise, interest in heritage has not only been unstoppable since its invention around 1830 but has continued to expand and encompass new objects.

2.2.3.2. *The quality of the place*

The quality of the place designates the specific dimension which grants recreation to the individuals, through the implementation of practices which fall under five categories: rest, discovery, play, sociability and shopping. But most tourists resort to combinations where one dominant category (in terms of time spent or preference) influences the choice of a destination. To satisfy tourists, who are after all, temporary inhabitants, the places were selected in function of the presence of the elements required for the accomplishment of recreation practices. They were also equipped with complementary activities at the initiative of endogenous or exogenous actors who identified tourists' needs. The association with spatial categories depends on the need (or not) for "given" or preexisting components[17].

17 We formally reject the term "natural", because while these elements are "given" in the sense that they are not created by man, they are appreciated by the human gaze. As a result, their meaning varies depending on individuals and societies.

Thus, recreation areas require either the presence of slopes, snow-covered in winter for skiing (but not in summer, for hiking and climbing) or an association between the sea and the winds for coastal forms of sliding. On the contrary, discovery is based on the diversion of elements already there. Castles, cathedrals, old neighborhoods or ancient villages were not designed for tourism, but the practice of sightseeing has taken over. In a similar manner, modern towers erected to accommodate offices, or even hotels, are only valuable for their aesthetic qualities or as platforms for contemplating panoramas. There is an opposition between the ancient margins (which were available and conquered by tourist stations and resorts, built ex nihilo) and the towns and villages (which belong to the full world inherited from different historical periods, and which are available for discovery). Later on, discovery practices also integrated the "monuments of nature", or breathtaking natural elements, elevated to the rank of monuments. This began with the invention of the natural park in the United States in 1873 (MIT team 2005) and expanded afterwards. The success was such that today many so-called natural, delimited and classified spaces are the subject of discovery, contemplation or journeying practices. As Table 2.3 shows, spatial technologies have been integrated to facilitate the accomplishment of tourists' plans.

	Spatial category	Given elements	Spatial technologies
Rest	1: Beach 2: Countryside	Sea, sun Calm	Stations and resorts Club stay
Discovery	Real world: City and sites Heritage and modernity Heritage campaign	Complementary (geographical site)	Accommodation Transport, museums Organized or self-organized tour
Recreation	1: Mountain 2: Coastline	Slope only or + snow in winter Sea + wind	Facilities Learning places
Shopping	Urbanity	None	Street or shopping center
Sociability	Ubiquitous	Ubiquitous	Walks, party places (casinos, night clubs, etc.)

Table 2.3. *Matrix of the quality of the place according to the representations of Western tourists. The numbers reflect a hierarchy in the practices of individuals*

The example of *trekking* illustrates how the combinations work. First of all, it is an experience that invites the body move. Indeed, it also incorporates elements of discovery through the contemplation of landscapes and the interaction with other types of society. In terms of representation, a destination first appeals to one or another dominant category, and on a second instance, to the locally possible combination. Thus, in addition to being defined by its perimeter, the destination has certain content. Over time, places become more and more sophisticated. The oldest tourist stations are full of features, villas, palaces, casinos, etc. suitable for satisfying discovery as a complement to the practice of rest, which is usually the dominant expectation.

The quality of one or more places within the destination, namely the capacity to meet tourists' expectations, will play a role in revitalizing the LTS. In the Pays de Monts LTS, the Saint-Jean-de-Monts station undoubtedly has a ripple effect, due to differences in attendance, the presence of facilities, notoriety and the capacity to organize events throughout the year. Not in vain, it has been doing so for around 20 years. To the north, Notre-Dame-de-Monts and Fromentine, the seaside district of La Barre-de-Monts, and to the south, Saint-Hilaire-de-Riez (too fragmented), cannot compete. The dynamism of the place is largely due to the structuring of the LTS. The preceding organization, the semi-public company, succeeded in bringing together the actors and promoting the quality of the place. A special emphasis was placed on welcoming families and on the great safety offered by a gently sloping beach, not subject to the dangerous phenomena of pools and tarpaulins (which form during the tides and trap unaware or poorly competent swimmers).

From this starting point, a range of adaptations have been made. For instance, workshops were organized during school holidays to welcome children. A kid's corner was set up within the tourist office. Totems are planted on the beach to give children directions, etc. But this polarization effect played by the station does not reach the islands of Noirmoutier or Yeu due to asserted local particularities, a rejection of the image transmitted by the station, as well as the distance and transshipment for the island of Yeu.

On the contrary, in other LTSs, operationality is limited by the absence of a ripple effect emanating from one or more hubs. Thus, the Saumurois does not live up to its name. While there is indeed a territory over which the town exerts an influence as consumption basin or living space, its capacity to generate ripple effects on tourism dynamics is weak. The castle, which occupies one of the most spectacular sites among those built along the Val in the 16th and 17th centuries, and whose physiognomy became known thanks to the *Très Riches Heures du duc de Berry*, has received little promotion for a long time and is not at all lively. The *Cadre Noir*, another well-known emblem, also favors the image of culture over that of tourism. More often on tour for demonstrations around the world than in the location, except for training, the local effects are measured. For a time, the hub's management was claimed by the Abbey of Fontevraud, which is not lacking in charm. But it is located on the outskirts, which is not an advantage. Above all, after the proactive leadership exercised by David Martin from 2007 to 2016, the place fell back into a certain apathy, alternating indecisions and windfalls (such as the installation of the happy donation from Martine and Léon Clingman, which certainly enriches the place, but will not be enough, for lack of sufficient determination). Through various vicissitudes, the territory has been trying to structure itself for 50 years, without really succeeding. As we saw in the case study, a reuniting structure was active in the 1990s, but was imposed from the outside (by the Regional Tourism Committee, within the framework of the state-region plan contract). No sooner had the pressure been released than the unit gave way to a mosaic of smaller entities (one around Doué-la-Fontaine, the other around Longué-Jumelles), thus reducing Saumur's area of influence.

In other cases, the ripple effect does not come about, due to the place's lack of sufficient quality. In Amnéville, tourismification owes much more to the will of Jean Kiffer, mayor from 1965 to 2011, than to the place itself, which, incidentally, was lacking in quality. For that matter, development was based on leisure, which means it had a spatially restricted aura. Furthermore, the death of its creator has left a void that the new municipality is still trying to fill. It seems this has not prevented a moderate success: the weakness in accommodation positively qualifies the place, which is reflected in the volume of jobs. In Lens, this weakness in the quality of the place, again associated with industrial history, constitutes a handicap, which is not found in Metz, where another relocation (that of the Pompidou center) was carried out in a completely different local context. The ripple effect is undoubtedly real.

Sometimes the ideal size is debated. Within Destination Anjou, some actors were in favor of an extension towards the east, Touraine, which could foreshadow an extended network in the Loire Valley. But others feared that this dimension may work to the advantage of operators located in the heart of the tourist region, rather than those of Maine-et-Loire, located on the margins.

2.2.4. *The different methods for inserting tourist places into the tourist lived space*

Tourists construct their lived space by constituting networks of places selected and visited according to their qualities. As individuals within the same society share values, the repetition of the same secondary movements endows these modalities with permanence and scope. A hierarchy is also established at a worldwide scale, shared in such a way that globality can be described as universal (Violier and Taunay 2020).

Several types of places can thus be identified depending on the position they occupy within these configurations. Based on an analysis of tourist practices in Gijón in Spain, Ballina (2021) has identified four.

– The "unique" location is sufficient to constitute the destination, including one or more nights. We can refine this type by distinguishing two subtypes. On the one hand, some are accessible, and located at a relatively short distance from the departure centers. On the other hand, metropolises are characterized by a high concentration of objects, standing out for their great interest. They justify a single trip with several overnight stays. Tour operator catalogs offer experiences in New York, Paris, London, Beijing, etc.

– The hub presupposes a main mobility and combines several excursions from a pole[18], understood as additional mobility carried out during the day, without an overnight stay. In that sense, Paris is a unique destination, but can become a hub as soon as tourists make secondary trips to Cheverny, the Loire castles such as Chambord or Chenonceau, etc. (as proposed by the main incoming agencies located in the capital).

– The stopover is a halt in a reticular itinerary including several places, separated by distances in such a way that the excursion seems less attractive. This choice depends partly on the operator and also on individuals. The most mobile ones will probably choose this form, whereas an older audience may prefer a polar system, which avoids switching accommodation and the difficulty of arranging and transporting luggage every day.

– A place can be a chain link within a network or a hub, and only be frequented during the day, without overnight stays on site.

– To this classification, we can add the scenario of a hub stopover (when a place is at the same time a stopover in a tour and part of a hub, meaning that several excursions are added to it). Thus, a metropolis like Paris is both a hub from which several day excursions are planned (such as the Loire castles or that of Versailles), and a stopover within a discovery tour which will also include the Giverny gardens and the Normandy coast.

In addition to the circumstances of time and place, or space, the strategies of the actors are relevant to understand the construction of Localized Tourism Systems.

---

18 In French, the term is ambiguous; it can mean a short trip from the main residence, as a trip from a vacation residence. We use it here in this second sense.

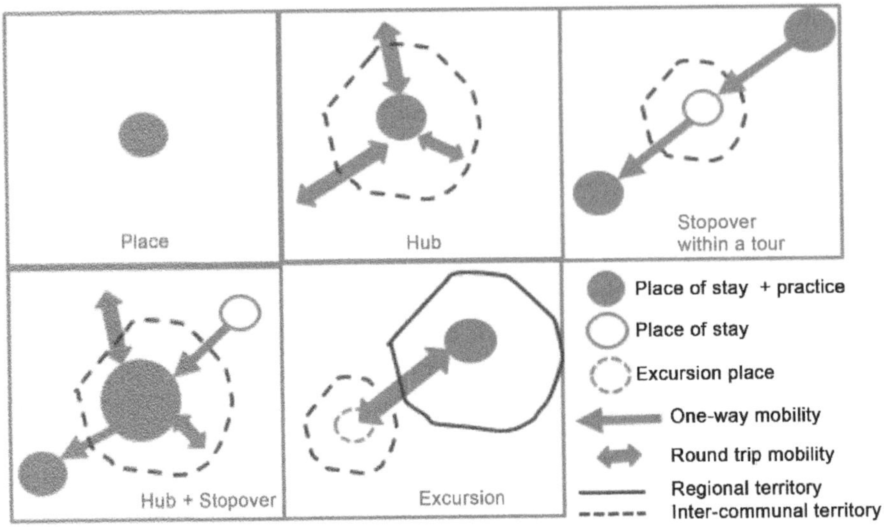

Figure 2.2. *Types of places according to tourists' attendance modalities, adapted from Ballina (2021). For a color version of this figure, see www.iste.co.uk/louart/touristdestinations.zip*

2.3. The games of actors

At the beginning of the place's tourismification, the tourist system was established through a meeting between tourists and residents. Places experienced development through the arrival of tourists, and the meeting of locals (sometimes fishermen, farmers) offering hospitality. As mentioned earlier, this is the case of Arcachon: in her 1977 thesis, Micheline Cassou-Mounat stated that this system foreshadowed the development of accommodation infrastructure or communication routes (Cassou-Mounat 1977). However, hardly had the tourist phenomenon increased when other actors intervened for regulatory purposes. Among them, states, companies and local regulation were implemented with the goal of establishing governance to ensure space promotion and management. Divisions of the tourist space were made, establishing scales of intervention (local, departmental, regional) that did not necessarily correspond to tourist practices. Tourists produce their own territory, in compliance with a recreational project and choose the places that justify their change of place. This division reminds us of what Claude Raffestin described as territorial imposition (Raffestin 1980), to the extent of transposing an administrative network that does not acknowledge the spatial practices of tourists. On this point, entrepreneurs and political actors are opposed in their relationship to the territory. For their part, entrepreneurs find themselves at ease with a reticular approach, as it brings forward the best skills or diverse clienteles, regardless of attachment territories. As for politicians, they favor the territory on which their legitimacy is based, in other words, the delimited space where they were elected.

Furthermore, at the local level, all of the actors of tourism do not share the same approach in their commitment to the places. Elected officials and technicians, who depend on public or para-public institutions (structures largely financed by public shareholders), seek to order a territory through spatial development and planning.

Other actors, notably entrepreneurs, operate through networks. Work on local productive systems has proven that spatial proximity among actors affects the functioning of local relationships. According to Grosseti (2004), "spatial proximity has effects on coordination between organizations, the probability of the existence of relationships increases as the distance is reduced" (Grosseti 2004, p. 167).

Yet, spatial proximity between actors does not constitute a guarantee of cooperation among them. Research on clusters frequently relies on the urban agglomeration as study scale (Bouba-Olga and Grossetti 2018). We understand that public institutions intervene to boost the economy and employment in these territories by creating dedicated spaces (technopoles, concerted development zones, etc.) and seek to stimulate cooperation between companies based on proximity.

According to Torre, local relations interpellate the spatial dimension, except in the analysis of "relationships between men and their activities in space, where space and geography count, but are not always decisive or favorable" (Torre 2016, p. 466). Applied to tourism, the dynamic of proximity is based on a supra-local and multi-scalar organization (Clergeau and Violier 2012). The ability of certain actors to act upon the functioning of the place makes it possible to reach a local consensus by uniting forces. Still, the divides between different cultures are multiple. But we will see that tourists are involved in the spatial delimitation of the system and contribute to creating proximity.

2.3.1. *Predominance of the tourism sector in the system*

2.3.1.1. *Trust between the actors in a sector and a territory*

In Saint-Jean-de-Monts, in a context where a crisis was glimpsed in the 1980s (due to the shrinkage of the trade area and the drop in the average spending per tourist), the creation of a mediating organization in July 1993 proved to be a solution. The mixed economy company brings together the multiple actors of the station around common projects: *marketing* positioning or even actions concerning the environment and sustainable development. The LTS is recognized as central, since in virtue of its status, it summons public and private actors, although the local community (in this case, the municipality), still holds the majority. Despite inevitable tensions, the structure has enabled mediation between socio-economic actors whose professional cultures may seem diverse, or even divergent. Notwithstanding the competence transfer to the inter-municipality in 2017, the federative dynamic will stand the test of time in Saint-Jean-de-Monts. The activities of the Saint-Jean-de-Monts semi-public company continue but are limited to a municipal territory. The local public company has an inter-communal territory. Respecting the law, a public company excludes tourism professionals from decisions, as this status allocates capital which is fully owned by public actors. However, *Océan-marais-de-Monts* (a tourism public company) has chosen to establish a technical committee of 25 professionals in order to ensure consultation and maintain certain legitimacy for a key actor.

This same trust relationship is found within the Lake Geneva basin, between the actors of tourism and leisure. To begin with, the managers of tourist sites (museums, theme parks, etc.) began to collaborate on the basis of the Swiss cantons (Valais and Fribourg). Then, the partnership extended to the French shore of Lake Geneva. This cooperation was born from the impossibility for these professionals to ensure the promotion and communication of their establishments, hampered by the administrative boundaries of their institutional territories. The birth of the *Léman Sans Frontière* association in 1995 aimed to lure clients into circulating first between the lake shores and the Pre-Alps, and then on both sides of the lake in France and Switzerland (Piriou 2018a). The members of the association shared a cross-border project to make their establishments known throughout the Lake Geneva region. This resulted in cooperation between a few sites around a common product, or in targeted communication operations. Later, the association sought to acquire legitimacy with the public authorities. Since the 2010s, the request from the Lake Geneva Council for European funding under the France–Switzerland cooperation has sought to perpetuate this system over time. It built a network deploying proposed combinations between tourist sites and itineraries within the cross-border space.

The role of politics may prove necessary for the success of such a system. It can act as a facilitator. In the case of Saint-Jean-de-Monts and *Léman Sans Frontière*, the role of politics was essential to support projects. Conversely, the lack of a willpower will lead to difficulties in articulating and making such a system work.

2.3.1.2. *The weight of politics*

The organization of a system by the actors of tourism in a favorable context may be subject to discontinuities and tensions. Political power can intervene to provide a vision and encourage collaboration. Despite this, the increased development of territorial *marketing* coupled with a political takeover of tourism-related subjects has led elected officials (and ultimately technicians) to define a destination that fits an imposed territory. Let us take the case of the Dordogne Valley. Located at the confluence of three departments, Dordogne and Corrèze (in the region of Nouvelle-Aquitaine) and Lot (in the Occitanie region), this tourist destination holds an exceptional monumental heritage, including the caves of Lascaux, picturesque villages, from Rocamadour to Eyzies-de-Tayac and Collonges-la-Rouge, small tourist towns like Salers, shared practices such as canoeing-kayaking down the rivers or exploring gastronomy. Despite the success obtained in bringing together municipalities and inter-municipalities throughout the north of the department of Lot and in the south of the department of Corrèze, the ambition did not succeed in integrating the south-east of the department of Dordogne (called the "Périgord Noir") into a single entity, the "Dordogne Valley". The personalities played a definitive role in engaging the actors and networking. Despite being politically opposed, two elected officials agreed on the challenges of tourism development in the north of the department of Lot. Then, they approached elected officials from the inter-communities in the south of Corrèze in order to join this inter-community, inter-departmental and inter-regional tourist office. Such approach was eased by the encouragement of other elected officials and socio-professional actors, *leaders* of opinion who followed this movement and helped to incite others to follow said approach (Piriou 2020b). However, this dynamic stopped at the gates of the department of Dordogne. Although the tourism technicians of the Périgord Noir and the Dordogne Valley reached a basic agreement for the sake of joint promotion, the elected officials of the department of Dordogne did not want the Périgord Noir tourist office to integrate the Dordogne Valley tourist office. The disparities in tourist intensity between the eastern part of Périgord, known as "black" (*noir*), and the rest of the departmental territory prevented an otherwise promising integration into a homogeneous whole reuniting forces distributed across several territories.

The role of politics can also seem intrusive for certain actors of tourism. In the department of Morbihan, the General Council's approach to operating a departmental-scale cluster proved to be a challenge, so much did interests diverge. The diversity of actors created dissonance regarding the goals of a federative project in the interest of a collective need. During the cluster's working sessions, while the owner of a castle in a small town in the center of the department desired to benefit from tourist flows (particularly in the summer), the head of an events facility at a larger coastal town was trying to deseasonalize its activities. The ambition of the cluster supported by the department was to encourage collaboration between the actors, especially those from the coast and the hinterland. The idea was not only to spread tourism throughout the territory but also to distribute its economic benefits. However, certain actors only became involved for information purposes, without any real need for action from the General Council. Given the weight of their structure within the tourism economy in the department, autonomous operation was favored. Throughout the working sessions, these same actors hardly became involved, or even challenged the public actor as to whether it was able to act within a time frame similar to that of private actors. The time taken by the public actor's decision-making processes, which depend on the vote of elected officials, turns out to be longer than the time used by private actors, who have to comply with a company's schedule.

We have shown that the double homogeneity of actors – their activity falls within a framework limited by a well-identified tourism sector, on the one hand, and they all belong to small and medium-sized businesses, on the other hand – facilitates the construction of a localized system and streamlines its operation, as evidenced by the case of Saint-Jean-de-Monts. However, this is not enough; the failure to give birth to a large "Dordogne Valley" illustrates this. What happens when a heterogeneous economic dimension reigns between the actors?

2.3.1.3. *On the proper use of the actors' economic heterogeneity*

The heterogeneity of actors and their different sizes contributes to creating tensions, disrupting the implementation of a system. For example, in the Loire Valley, castles that experience significant attendance (close to a million visitors per year) have been slow to cooperate with others with lower attendance. It was not until 2009 that the Royal Valley association (created in 1979 and bringing together the largest), joined the historic Ladies of Touraine Route (born in 1975 at the instigation of the owner of the castle of Montpoupon). The Loire castles association then reunited 57 sites upstream to downstream of the Loire Valley, between the Briare canal and the estuary. In defiance of what historians describe as the "Loire castles", the ensemble also brings together a relatively homogeneous array, dating from the Renaissance. The goal of the Loire Castles-Royal Valley (*Châteaux de la Loire-Vallée des Rois*) association is to generate exchanges between members and ensure common promotion. The ambition is also to increase the visibility of the lesser-known itineraries, thanks to networking and through the intermediary of the most emblematic castles. Upon its creation, however, some castles (*châteaux*) remained reluctant to this cooperation. This was the case of Chenonceau, which at the time opted for independent promotion and marketing. At present, this castle adheres to it. However, a network of 23 "major sites of excellence of the Loire castles" was referenced in a charter of excellence for international marketing purposes. This approach included promotional actions, the improvement of mediation, quality adaptation and network coordination, led by the regional tourism committees, the Mission Val de Loire and the Departmental Tourism Committee in Indre-et-Loire. However, the "Châteaux de la Loire" (Loire castles) brand was not registered with the INPI when it was created in 2007, as it happened with other brands like "la Loire à vélo", whose owners are the Regions of Pays de la Loire and Centre-Val de Loire.

Only a graphic identity was used in the design of the "Châteaux de la Loire, *finest France*" (www.loire-chateaux.org) website, hosted by Atout France. Since the admission of Montsoreau Castle in December 2020, this network no longer seems to be really active, at least in terms of promotion, since the dedicated web page no longer exists. As for the "Châteaux de la Loire-Vallée des Rois" association, it had 103 castles in 2023, and wished to promote all of the member castles to the general public. Tourist institutions are partners: tourist offices, private companies (banks, auctioneers, etc.) and the two regional councils (Pays de la Loire and Centre-Val de Loire). The dynamic driven by private actors seems to persist in the face of the trajectory of public institutions.

The homogeneity in the types of actors facilitates networking and the functioning of the system, but is heterogeneity always an obstacle? In the case of Saint-Jean-de-Monts, despite the homogeneity of small independent companies, we note the presence of a few establishments having adhered to national and international networks. Nevertheless, this only slightly affects the dynamics. Some campsites managed by the Siblu group ostensibly benefit from being integrated into a local ecosystem. The interest for this brand is to ensure cooperation within the tourist destination so as not to be marginalized within a territory where competition between campsites is strong. Integration also grants participation at events organized by the institutional structure, offering the company the opportunity to promote mobile homes (whose sale constitutes one of the economic drivers of the group). Its economic model is based on the sale to individuals, who, after having installed the equipment on their land, offer it for rental, either by Siblu, by individuals who go through intermediaries on the Internet (Leboncoin website) or specialized micro-societies. The latter, who also provide concierge services (the reception and departure of tenants, as well as cleaning) and usual formalities (signing the lease, remittance of payments, etc.), flourished around 10 years ago, helping to enrich the local system of actors.

It may appear that heterogeneity in the types of actors can be beneficial (and even desirable) to strengthen the system. Well-known actors encourage tourists to come to the region, while the most modest ones benefit from such windfall. The latter then contribute to increasing the stay length, travelers staying longer and avoiding excessively long wanderings. The Futuroscope Park, having opened north of Poitiers in 1987, has been a locomotive for many small tourist sites in the Vienne department. The *Vallée des Singes*, opened in 1998 in Romagne, or the Planet of the Crocodiles, inaugurated in Civaux in 2008, benefit from the notoriety of the Futuroscope Park, as well as some visitor flows. The department's communication team does not hesitate to use the expression "the country of Futuroscope". In the Lake Geneva basin, heterogeneity in the types of actors is also profitable, since tourist sites with a large number of visitors make it possible to publicize sites with less frequent visitors, by distributing the *Léman Sans Frontière* map and the guide available at the site's reception. Within this destination, tourists travel through the various cities and attend the different sites.

As long as the actors play in the same league, proximity facilitates professional relations. Economic heterogeneity can become a lever to energize the whole, and in special, benefit the smaller structures, who do not always have the means to increase their influence. But a second context can also be found. It is marked by the attempted incursions from other economic sectors, seeking an answer to specific issues in tourism.

2.3.2. *Tourism associated with other sectors in the system*

Throughout its history, tourism has often developed in correlation with other economic sectors (e.g. agriculture). The Alsace wine route, created in 1921, helped the vineyard's winegrowers not only to promote their products but also to diversify their activity. Yet, the rapprochement of two sectors is not obvious. To create a system, the two sectors must agree on shared goals and purposes and recognize legitimate piloting.

In Nantes, while the establishment of a localized system of actors marrying the spheres of culture and tourism experienced a complex onset until 2012, the implementation of a new structure (the public local company *Voyage à Nantes* (VAN) made it possible to rationalize the strategies identified by the multiple actors. Beyond the grouping of structures (the semi-public company Nantes culture heritage, responsible for the *Machines de l'île* and the castle, the associative tourist office, the semi-public company of the Congress center or the *Estuaire* biennial, managed by the *Lieu unique*), the extension of the system was remarkable for the opening to related territories in the capital of the new local public company *Voyage à Nantes*. The towns of Saint-Nazaire, the inter-communalities Estuaire and Sillon, Sud Estuaire, Clisson, Sèvre and Maine Agglo also financially contributed 2–3% to the *Voyage à Nantes*. Described as "branches" by Jean Blaise, its director, there was an extension of the territory towards the estuary (west of the town) and towards the vineyard (south of the town). This perimeter introduced events and artistic equipment deployed along the peripheral municipalities. At the end of the 2000s, Nantes was not part of tourist itineraries, the city did not constitute an obvious gateway (Violier 2009), and yet the supply now has a strong influence around the city: "Nantes, starting point for walks" towards the Loire, towards the estuary, the vineyard or even towards the Nantes-Brest canal (tourist guide "Nantes le voyage permanent", 2020–2021, p. 141). Through the "modern crossing of an old country" project, Nantes was set against a larger tourist network and positioned at the center of an axis ranging from the Loire Valley to Mont-Saint-Michel.

In the case of the Autour du Louvre-Lens (ALL) destination contract, the aim was to strengthen and support the first connections between the tourism and the culture sectors, notably driven by the listing of the mining basin under the UNESCO World Heritage. Tourist development especially relied on the supply representative of the territory's identity (slag heaps, museums, cultural routes). However, political reactions arose as to whether the cultural orientation relating to local development should be reinforced. A remark was also made on the lack of involvement from the local population due to the significance of the given cultural orientation: a relocation of an elitist Parisian model to the detriment of more popular local cultures.

In the Cognac vineyard, companies and industries related to the spirits sector (glassworks, cooperages, etc.) were set around the town of Cognac in order to meet the needs of trading houses. Networks were built, both in the Cognac sector and in tourism, with the coexistence of two positions. On the one hand, the Cognac industry sought to promote its brands and generate bottle sales, almost entirely for export. Tourist tours from Bordeaux to the Cognac vineyard were offered by incoming tour operators (Piriou 2018b). In order to maintain those flows, in 2006, the department of Charente developed a partnership with the departments of Charente-Maritime and Gironde, aiming to ensure the promotion of a "Bordeaux-Cognac" destination, reframed by these same departmental councils in 2008 in the Charentais departments under the "Cognac Experience" communication. A desire to refocus a Charente identity around Cognac was established so as to break free from the sphere of influence of Bordeaux. This approach also earned them a destination contract in 2022. The strategy strengthening links between the two departments along the Charente River was marked in 2017 by the merger of the two departmental associations responsible for tourism development, Charente and Charente-Maritime, which came together under "Charentes tourism", as decided by the presidents of departmental councils. We may observe that the Cognac tourist territory managed by tourist institutions has gradually shifted into an administrative territory, reinforcing the links between the two departments along the Charente River, but independently from the Bordeaux tourist flow.

In Saumur, the International Tourist Hub (ITH) was an example of a unifying structure for the promotion of the territory. However, the match between the tourism sector and the equestrian sector seems imperfect. The Saumur Equestrian Committee was presented as the bridge for valorizing the equestrian sector and for solving the territorial challenges of tourism within the ITH. A study carried out among tourists has shown that the town of Saumur is not sufficiently perceived by its "horse" component, and there is a lack of knowledge of equestrian activities, something essential for the recognition of an equine tourism cluster (Clergeau et al. 2015). The horse gradually lost its economic usefulness, in transport as in agriculture, and even in the army. Leisure and tourism now constitute the fields of its reinvention, as shown by the work of Sylvine Pickel Chevalier (2019, 2020).

Developing a system can prove difficult, as the sectors the actors belong to are diverse and varied. Connections require not only the understanding and articulation of respective operations, but also the adaptation of a shared project. The distance between the sectors and the varied goals make this exercise tedious. This difficulty in articulating tourism with other sectors is found in many territories. The quality of intermediation would be a crucial determinant for the system's proper functioning.

2.3.3. *Intermediation as the key to the system's success*

Intermediation to ensure the system's functioning and coherence is necessary to define the collective interest. The role of intermediation involves connecting actors and circulating information between them, leading a network of actors, translating and interpreting the proposals from actors and coordinating a common project (Clergeau and Violier 2012). We can identify two types of intermediations: controlled intermediation and negotiated intermediation.

2.3.3.1. *Controlled intermediation of the system*

The personality and audacity of the "author" actor could contribute to the system's sustainability. In Nantes, the project to reconvert the island – a disused industrial district – was the result of a study commissioned by the city's elected officials in 2001. Several proposals were brought forward with the idea of developing a creative metropolis. Some aimed to transmit the memory of places through the exploitation of museums. Others were based on the culture, technique, and know-how of the naval industry. Finally, some creative projects emerged, including the *Machines*, by the Manaus association, under the leadership of Pierre Orefice and François Delarozière. But none of the proposals would immediately correspond to the research firm's intention of identifying a project for developing a major cultural and tourist facility. The *Machines* project was even poorly evaluated. However, Nantes Métropole asked another firm specializing in tourism consulting and engineering to carry out a second study on the feasibility of the *Machines* project (Barthel 2009). The role of the city's deputy mayor, Jean-Marc Ayrault, is clear and intends to leave his political mark on the territory, especially since he was elected president of Nantes Métropole in 2001. In the shadow of the elected official was Jean Blaise, the architect of Nantes cultural policy. This man had been active since the first municipal mandates of Jean-Marc Ayrault and directed numerous projects (festivals, transformation of places, etc.). Apart from the *Machines de l'île* project, which saw the light of day with the presentation of the great elephant in June 2007, there was the rapprochement of Nantes' institutional culture and tourism spheres, materialized by the creation of the *Voyage à Nantes*. This confirmed the audacity of a political/artistic duo who successfully set up a tourist and cultural system in Nantes (Morice and Violier 2009).

But was that boldness sustainable? In the case of Nantes, to this day, the question of succession is open. Jean-Marc Ayrault, engaged in the role of Prime Minister in 2012, ceded his mandate as mayor to his first deputy, until 2014, when municipal elections were organized, and Johanna Rolland was elected. However, Jean Blaise has remained as the director of *Voyage à Nantes* since 2011.

On the Amnéville side, succession was the source of numerous difficulties. The mayor, Dr. Kiffer, who instituted a conglomeration of establishments, experienced limitations both on the organizational and the competitive levels near the end of his life. In May 2010, the creation of a Center Parcs for leisure purposes approximately a hundred kilometers away, that is, within the same trade area (the SaarLorLux euroregion with over 11 million inhabitants), contributed to the need to review the development model. Furthermore, financial difficulties and administrative burdens compelled the ensemble to evolve towards more autonomous structures, at the risk of integrating external investors with little interest in collective initiatives in favor of the development of Amnéville.

In the department of Morbihan, the departmental council wanted the Departmental Tourism Committee to relinquish its role as network unifier and coordinator. Rejection was expressed by many professionals who were used to working with the tourist institution. In addition, several of those professionals collaborated informally on different projects (sustainable development, group reception). The mediation imposed by the departmental council enabled actors who had not been included in collaborative networks to engage in a new collective dynamic supported by the department.

Intermediation is necessary to manage the relationships between the actors in an LTS. It makes it possible to smooth out the goals and purposes, as well as to harmonize the heterogeneity of the actors. However, the legitimacy of the actor responsible for intermediation determines the evolution, even the trajectory of the system. In the event of rejection, the network effect powered by the intermediation actor diminishes in favor of more restricted, preexistent informal forms of collaboration. Negotiated intermediation differs on this point.

2.3.3.2. *A negotiated intermediation of the system*

The system's negotiated intermediation operates according to a more community-based approach. The choice of the actor responsible for intermediation is up to the actors who are members of the system and establishes partnership intermediation (which involves part or all of the actors together).

Destination Anjou, reuniting hoteliers, restaurateurs and managers of leisure facilities, is managed by the Maine-et-Loire Chamber of Commerce and Industry. This body is recognized as legitimate for supporting local businesses to increase their performance and to cooperate by means of customer referrals. The homogeneity of the actors (mainly from the hotel, catering and leisure sites sector) facilitates exchanges, not only through informal relationships, but also through partnership mediation orchestrated by an intermediary, representative of the sector.

This negotiated intermediation approach is also present in the *Léman Sans Frontière* association. Certain members constitute the office and the association, while the general secretary has an intermediation role. Membership to the association, featuring a volunteering role, reflects the choice of the actors to collaborate, be them French or Swiss (Piriou 2018a). It operates as a system insofar as the members ensure promotional links between two or more establishments and visitors combine visiting locations. The impetus initially given by a few sites, as a result of the lack of promotion from institutional tourism bodies, was also supported by French local authorities and Swiss public institutions a decade later, obtaining European funding to facilitate cross-border exchanges.

Is intermediation nonetheless essential? The Mélusine association in the south of Vendée demonstrates that collaboration between heterogeneous actors (restaurants, hotels, leisure sites) creates a homogeneous system with the goal of promoting the territory and ensuring links to encourage the arrival of visitors. However, there is no intermediary actor. There is a functional complementarity between them in order to offer a global supply for the visitors' stay, so it may seem that the system works by itself, without any additional actor.

2.3.4. *The spatial trajectories of systems*

Each system has an often local history related to a problem. Networking makes it possible to compensate for the functioning of the territory deemed unsatisfactory (or even failing), regarding promotion or the reception of visitors. Finally, we observe two forms of development at the spatial level: first, the birth of a local system that extends to a regional level. Then, a system restricted to the local sphere or even imposing itself within administrative boundaries.

2.3.4.1. *From the local demand to expansion*

A critical situation or the observation of a dysfunction is often at the origin of actor networking. If relationships are established and function properly, then a localized system is born. As a result of success, the system's spatial dimension tends to increase, allowing for the integration of members who also correspond to a space apt for tourist practices.

In Nantes, in 2011, the project bringing together the culture and tourism sectors resulted in the creation of the local public company *Voyage à Nantes*. After 2015, the actors from Nantes reached out beyond the metropolis and joined the actors from the Nantes vineyard, among other places, around the city of Clisson. Since 2007, the Estuaire biennial has offered an exhibition of contemporary works along the Loire, from Nantes to Saint-Nazaire. This activity is part of the local public company *Voyage à Nantes*.

In the case of the *Léman Sans Frontière* association, the collaboration was born between two places: the pre-Alpine Mountain resort Moléson-sur-Gruyères and the Swiss Vapeur Parc theme park, located in Bouveret, on the eastern shore of Lake Geneva. Then, other members around the lake joined forces: the *Compagnie Générale de Navigation*, based near Lausanne (Switzerland) and the *Jardin des Cinq Sens*, in Yvoire. The association reached the number of 41 members on both sides of the Franco-Swiss border.

An intensification dynamic of the LTS can also take place through the integration of new actors (either formerly excluded or recently emerging). In 2022, the public company of the CM *Océan-Marais de Monts* decided to welcome the owners of furnished accommodation for short-term rentals.

2.3.4.2. *Local restriction of the system's spatial dimension*

The conditions for the emergence of the system account for the possibility of its deployment in space. The actor or actors who drive the networking contribute to choices regarding the system's spatial inscription. The choice may involve trying to develop an LTS within an administrative territory, while respecting geographical boundaries.

Destination Anjou is an association created on the initiative of hotel, catering and leisure facilities providers in the department of Maine-et-Loire. The proposed offer is also structured to ensure the promotion of the department. The privilege passport box in Anjou was developed with this in mind, even though this project did not have the hoped-for success. Since Destination Anjou is hosted by the Chamber of Commerce and Industry of Maine-et-Loire as part of its quality mission, extending the spatial dimension of the system is not conceivable.

In Morbihan, the desire of the departmental council to pilot the tourism cluster project immediately restricted the system to the department's boundaries. A form of territorial imposition was implemented as the political actor wished to foster exchanges only between the actors from the same department. Certain actors, from the Brocéliande forest in the north of the department, close to the Department of Ille-et-Vilaine, were used to cooperating with their neighbors. However, within the framework of the cluster, they were unable to invite these same neighboring actors with whom they were used to working on a regular basis, since they were not located within the department of Morbihan. The system's spatial dimension was restricted, and upon being confronted with this territorial imposition, the relationships between the actors were reduced to informal contacts preexisting the clustering process.

3

Managerial Analyses

In this chapter, we seek to understand "Localized Tourism Systems[1]" from the viewpoint of management sciences.

Section 3.1 will describe the common points to all of these systems (territories functioning within the frame of "destination contracts", clusters or other tourist configurations studied in this work).

Section 3.2 will show how the highly variable forms that these systems take make it necessary to analyze them differentially. For that, we should characterize them not only in function of their history and territorialization but also their resources, projects (implicit or expressed purposes), the way in which their actors interact (through networks, complementarities, rivalries or collective development capacities) and, finally, their forms of governance.

Sections 3.3, 3.4, 3.5 and 3.6 will specifically describe what we mean by resources, projects, networks and governance, applying these concepts to Localized Tourism Systems.

---

1 We chose this expression to illustrate the fact that tourist activities are territorialized ("localized") and constitute systems of actors with joint interactions and regulations, having more or less elaborate collective goals (see further development in the text). We have not employed the term in vogue, "ecosystems" (as used by Laperche et al. (2019)), because we believe it should be reserved for systems studied from an ecological perspective, as competitive spaces where the place's equilibrium prevails over the marginal interest of a category of actors. For example, we can talk about the ecosystem of Saint-Jean-de-Monts (including nature, animals, plants and humans, but excluding tourism) and find, within it, a Localized Tourism System. We could still speak of a tourism ecosystem if we referred to the overall interactions within a tourist territory, including the effects of tourism (and its actors) on every observable interaction (ecological environment, human and nonhuman living groups, or even technical, material or climatic artifacts with various influences).

We will refer to the cases evoked at the beginning of the book to demonstrate our analyses and show how to interpret the ample field studied ("tourist destinations as systems"), from an operational perspective. Each of the chapter's sections and subsections corresponds to a theorized presentation, in parallel to the aforementioned cases[2].

Chapter 3 intends neither to classify those cases nor to define the problems related to recommendations (remediation, how to profit from opportunities, development options). Instead, this will be the subject of Chapter 4.

3.1. How to manage tourism systems?

How are managerial studies capable of analyzing a territory and its economic (in this case, touristic) potential? To do so, we can find inspiration in several theoretical models[3].

3.1.1. *Systems' characteristics*

First, when we talk about a system, it is imperative to correctly grasp the concept. Without necessarily tracing the works of Tabatoni[4] or Le Moigne, a decent summary has been established by Leloup (2010):

– the idea of a system responds to the logic of an active totality (a whole which is at the same time more and less than its parts), with a purpose (explicit or not), internal actors (exercising interdependence and interaction) and an evolutionary dynamic (assuming self-organization);

– this corresponds to the current conception of territories in the political sense (where "Localized Tourism System" are born and developed). Endogenous collective actors (supposed to implement governance appropriate to their needs) express themselves so as to facilitate the internal integration or productive compatibility of their components, while establishing satisfactory contractual relationships with their environment.

---

2 With some exceptions, in this chapter, we have not redescribed (not even briefly), the cases, because these are detailed in Chapter 1, and for the most part, summarized for illustration in Chapter 2.

3 We use "model" in the sense of Le Moigne (1990, p. 5): "this makes a phenomenon otherwise perceived as complex intelligible, thus anticipating the consequences of possible actions within the phenomenon considered".

4 R. Perez, 2007, "Pierre Tabatoni et le *management*". *RFG*, 11(180), 15–24.

– In concrete terms, if we apply this model to tourism systems, this implies at least four types of complementary actions (described hereafter). It is therefore appropriate to:

1) design a collective project[5] (or, if one already exists, to reorient, revitalize or consolidate it);

2) improve available resources and structure their provision (by revitalizing the corresponding territory);

3) coordinate actors (by strengthening their relative cooperation and preventing them from entering into harmful tensions or destructive conflicts);

4) establish collective governance[6] suiting both actors and project.

3.1.2. *Tourist systems, their territory and temporality*

3.1.2.1. *Territorialization*

Localized Tourism Systems participate in the development of a territory based on tourism. The interactions taking place therein organize productive compromises between the available resources, the actors present and the resulting projects, not only with stimulating effects but also with tensions and gaps which have to be regulated (hence, the need for governance).

Systems are conditioned by historical and institutional temporalities that determine the practices of actors in thriving tourist places. The resources available, tourists' expectations and the terms of the supply depend on what is concretely possible at a given moment.

Tourist resources are socially construed, depending on their symbolic interest and practical arrangements. Take the example of the caves now available for visits (natural, painted or historically inhabited): their transformation into tourist resources has depended as much on a trend established by speleological discoveries (or an interest in rock art), as on access techniques having notoriously evolved over time.

5 What we call a "project" is a general orientation successfully contributing dynamism and a desire for action. It may be a collection of measures deemed useful and offering fresh impetus. Furthermore, project, resources, coordination and governance are developed interactively by mutually feeding their progression.

6 In *management*, governance is a flexible, participatory and concerted mode of direction. It comes from the same Greek root as "government" and "helm", initially associated with ship steering and orienting.

The same can be said about ski slopes and sea bathing[7], the proliferation of zoos[8] and other aquariums, or the recent installation of *escape games*.

In practical terms, resource management (their construction, the measurement of their insufficiencies or their potential, their use as tourist products and sources of income) depends on the actors involved and their capacity to plan the future together, to make choices (regarding supply and marketing), to boost collective action (through consultation and cooperation networks) and to accept a minimum of governance which will make the destination a recognized and ravishing place.

As will be shown throughout the chapter, this force of collective action also requires us to:

– reduce active internal conflict (of interests, objectives, people) to an acceptable level, so as not to block necessary collaborations;

– reduce the tendency to deviate from the tourist system we belong to (either because its usefulness has become incomprehensible, or because we are hostile to its main actors).

3.1.2.2. *Temporalities*

Tourism systems evolve over time (notably in relation to their resources, their actors and the institutional or economic contexts which influence them).

To interpret the temporal dynamics, let us refer to the model developed by Mendez (2010), according to which developments follow one another sequentially, in a chain of successive phases. Each phase shows a certain continuity in the combination between actors and contexts.

On certain occasions, a phase will be disrupted by one or more generators of change ("drivers", according to Van de Ven and Poole 1995): regulatory or institutional modifications (e.g. the NOTRe law); new projects or action programs within the territory; the disappearance or arrival of key figures; struggles between certain economic or political actors leading to transformative tensions, etc.

7 As already commented in Chapter 2.

8 Note that the symbolism of zoos has also evolved over time: "resources" today are not the same as in the middle of the 20th century. Animals currently have better equipped enclosures (removing their repressive aspect) and the duty to protect species is becoming a major issue; profitability nonetheless continues to be pursued by all means (catering, sale of objects, shows, subscription incentives, etc.).

The influence of drivers can trigger phase changes. This is often accompanied by bifurcations of trajectories, with systems potentially evolving towards goals or organizational modalities which diverge from the preceding period.

3.1.3. *Application to the cases observed*

We have hypothesized that the tourism organizations studied in this work function as systems, with interactions involving a minimum of overall coordination.

This is true for a certain number of territories. But in some cases, we should rather speak of conglomerates of activities. There is a juxtaposition of economic or political actors with weak ties and episodic relationships, as in Amnéville's leisure center or the *Saumurois*.

And we can also find combinations of several tourist systems in cohabitation, with partial cooperation. This is observed in the *Autour du Louvre-Lens* (ALL) destination contract, which combines cultural activities (museums), places of remembrance, revamped former industrial sites and sporting or leisure activities.

Historically, almost all of the systems studied have experienced differentiated phases, based on contextual drivers of change. Some of these drivers are disruptive, for example: the disappearance or arrival of an elected official or influential leader, as was the case for Amnéville or Nantes; the establishment of a new resource with significant impact, such as arrival of the Louvre in Lens; the effects of a sudden loss of customers, due to the Covid-19 pandemic, just to mention a few[9]. Other drivers are more progressive and act gently, such as the signing of a "destination contract", which can provide resources, develop programs, introduce actors or enrich the goals collectively set (Dordogne Valley, Louvre-Lens).

It is in function of this evolution over time that projects are implemented and whose details will be described below. They appear at the start of each new phase, encouraging or forcing them to modify their characteristics, their functioning or their goals.

At birth, each tourism system originates from a founding project:

– In certain cases, local actors seize opportunities in the territory. They implement activities which evolve through practice, without a formalized

9 An increase in customer numbers can also force strategic changes, with a consequent break in temporal sequences. In 2023, the small municipality of Île-d'Aix, overcrowded in the summer, was looking for ways to space out visits (as certain popular zoos or museums do, with daily quotas and requiring visitors to book their places in advance).

framework. Their project is emergent and implicit. This is evident in Tourism Innovation Lab in Angers.

– In other situations, development strategies are pondered, but without overly prefiguring a specific outcome. Projects are described and activated in a flexible manner (*Voyage à Nantes*, Dordogne Valley).

– We can rely on formal models, inspired in systems or schemes created elsewhere, or having been forged intellectually by ourselves, with action programs guided by political declarations or idealized scenarios (e.g. "transforming Lens into a city like Bilbao in Spain", developing the Languedoc coast by emulating other Mediterranean achievements, etc.). The projects are then tightly structured, with a preestablished goal (Center Parcs).

3.2. How to manage existing resources and enrich them or recompose them in an effective manner?

While the concept of system possesses general properties, the way in which each concrete scenario will evolve depends on local configurations and the actors at play.

Before approaching the actors and their productive combinations, prior to assessing the value of the projects they share and the manner in which their interactions are governed, it is important to analyze the products and services which constitute the tourism supply. Because this is what can be improved or developed, in connection with the tourists themselves and the interest they find in contemplating said supply as a tourist destination (stopover, prolonged stay, vacation, etc.).

3.2.1. *Resource-based management*

From a strictly economic point of view, our analytical framework is supported by "resource-based management" theory (Penrose 1959; Barney et al. 2001). It postulates that a substantial part of an organization's (or group of companies) success is derived from the resources it is capable of mobilizing. In relation to tourism, there is a double movement:

1) stimulating local cooperation;

2) using it to increase overall competitiveness in relation to competitors. In this sense, the economic perspective provides concrete goals (in terms of productivity or result expectations), without anything being removed from the above-mentioned systemic approach.

In parallel, we must rely on customers and prospects[10] (in other words, on the interest of tourists in the destination). With this question, we can resort to "open innovation" models, which integrate the reactions and proposals of companies' customers for enriching the supply of products and services.

But simple competitive mimicry does not suffice, which can result in a loss of strategic differential[11].

3.2.2. *The transition from current or potential availabilities to their representation as usable resources*

In a particular territory, we may find different types of local benefits:

1) Natural objects: landscapes, climate, heritage, traces left by history, urban development, artisanal production, sports equipment, leisure industries, etc. We can add opportunities in terms of men and financial resources. At the origin of Amnéville, for example, there was "the geographical location, the site, land availability, the presence of private entrepreneurs, the use of funds intended for the reconversion of Lorraine".

2) Events that occurred at a specific time or within a particular period. They have led certain actors to act, to profit from said advantages, to optimize them, to respond to financing or management problems related to territorial action political models.

These structural advantages are combined with trend opportunities (related to current mentalities, business tendencies, the supposed evolution of tourism, emerging or developing ideological movements).

---

10 When building a "destination", we can put an emphasis on supply (actors of the local system) or on demand (the interest of potential tourists for this destination, as well as the expectations expressed therein, both realistic and imaginative). Recently, while tourism specialists have emphasized the impact of tourists (without them, there would be no "destination"), economic analysis has shown movements in both directions. Sometimes it is the supply that pulls the demand, whereas sometimes it happens the other way round.

11 To avoid the standardization of tourist proposals, the territories have an interest in claiming their specificity, even though this means staging their "places" or their "history" (as Puy du Fou has smartly done). Moreover, histories are also geographies (the economic barge trips on the Loire, the erection of castles at key periods, the *terroirs* displaying their vines and the wars their vestiges, the industrial breakthroughs leaving their wastelands scattered). At their level, tourists highlight what attracts them in these representations, which acts as added value or differential advantage.

Advantages and opportunities become resources when they are mobilized for tourism action, which requires socially transforming them and making them have an active representation, for example, by endowing them with usefulness or attractiveness. Discernment about "tourist resources" may be derived from political intuitions, strategic choices, prospective visions, or the seizing of opportunities by local entrepreneurs, etc. We may also resort to in-depth studies on the evolution of markets or reference environments.

In the context of a systemic tourist destination, we may clearly see the interest in:

– understanding the resources available;

– increasing their strength or potential;

– activating those which are in an embryonic state, still under development or in insufficient amounts;

– relating these resources to probable changes in environments (in terms of opportunities or risks) because this is what will enable them to be linked to reality, to support (or reinforce) them, to make them relevant and to optimize their use.

3.2.3. *From resources to the tourist attractions' supply*

In practice, the basic resources of a tourism system correspond to all of the attractions available to visitors. These are the previously mentioned assets (landscapes, weather, monuments, cultural, sporting, recreational, exploratory amenities), to which are added access facilities, means of transport, accommodation or catering services.

These attractions [12] have to be sufficient in number, accessible and complementary. Their design should match tourist expectations and make people want to use them.

First-level resources (products and services) are not only linked to financial resources (which enable maintenance, operation and development), but also to human resources (effective organization, involvement, skills, collective intelligence).

---

12 The term "attraction" initially designated the tendency of certain physical bodies to approach one another (Newton's universal gravitation, molecular force of attraction in chemistry). In a figurative sense, in particular for what we call "tourist attractions", it shows that the attractiveness of a place or a scheme results from an interaction between proposals (on the part of professionals) and expectations (from tourists), following a mutual reinforcement logic.

3.2.4. *Public/private cooperation*

To the preceding elements, we should perhaps add a reality that can manifest as practical cooperation or tension.

Since they interfere with territories under public governance, tourism systems involve forms of cooperation between public bodies and private actors. These two types of partners have a compelling character and hold an interest. Institutional aspects and operational practices are mixed.

Within a territory, there is always a mixture of public and private resources. In general, monuments, natural curiosities, major facilities and infrastructures are managed by the public bodies; most attractions, small museums, parks, accommodation and catering services, by private companies, while the means of transport combine a mixture of public and private initiatives.

Most often, the public bodies retain overall authority. Their means are numerous: issuing authorizations, providing subsidies, infrastructure management, local government functions, etc. Developing and managing this type of cooperation requires complex forms of transactions, especially since there are often overlapping instances in public matters and conflicts of actors in private matters.

In all of the cases studied here, public/private cooperation is found. Their modalities vary from one situation to another. Note that this is the very essence of destination contracts.

3.2.5. *Between accelerated competition (with competitive imitations) and the need to offer territorial specificities*

Understandably, the inventory and management of tourism resources depends on observable trends in the supply and demand markets in this sector. For each territory, compromises or complementarities must be found between the following.

– The appropriate responses to competition, which has accelerated worldwide due to supply multiplication and media inflation[13]. We could have thought that this would be compensated by a parallel increase in demand, but there are doubts about this, at least regarding its importance. Because international mobility risks being reduced due to the high-energy cost of transport, to which are added the risks of terrorism or the pandemic, or even the hostility of the hosting territory's inhabitants, as soon as they perceive the tourist influx to be excessive.

---

13 Let us not forget the growing impact of influencers, whose experiences posted on social networks (Instagram, etc.) play a driving role in the attractiveness of places and tourist sites.

– The search for a multi-activity supply, with more cultural and sporting opportunities (inviting large and more or less prestigious events, depending on the volume of tourists in the sites concerned).

– Contributions connected to digitalization and the solutions it provides: on the publicizing and marketing of tourist products as well as on the content of the proposal made to visitors (through the multisensory and immersive experiences of all kinds that digital technology now permits[14]).

– The need to build tourist supplies closer to ecological requirements (new soft mobility in the territories, the search for attractions and more frugal reception solutions), or more in line with new social needs (contacts with residents, proximity exchanges, creative or sporting activities shifting away from passive tourism).

In geography (Chapter 2), there is major emphasis on the quality of a place, and beyond that, of a wider space combining tourism areas and itineraries, in order to produce attractiveness or meet the tourists' expectations for a destination. In *management*, we pay attention to the operational development of resources. Not only are there opportunities inherent in the tourist space (tourism area, groups of places, circulation) but also in the way in which its representations are deployed (images, stories, discursive or scenaristic focus), and how the reception and uses are organized (e.g. by facilitating relaxation, pleasures or distractions).

Tourist objects must always be evaluated in terms of affordance: they have intrinsic (potential) qualities and these are made perceptible to the ones who will benefit from them (attractiveness).

From this first description of tourism systems and their resources, we will now address three complementary topics:

– through the implementation of collective projects (section 3.3), associated with resource mobilization, we can develop the territory's allure (section 3.3.1) and increase the effectiveness of its supply (section 3.3.2);

– by optimizing the networks of actors (private, public, inside tourist structures and between them), we can strengthen the tourism system's potential (whether it is concentrated on a site or in a diffuse territory, whether there are strong convergences or competitive tensions) (section 3.4);

14 On these points and what follows, we resorted to prospective work by Welcome City Lab and its annual trend book, published with the support of *Atout France*. In notebook No. 7, we found a maxim by Henry Miller (frequent traveler), who skillfully insisted on how tourist resources are shared representations between suppliers and customers: "A destination is never a place, but a way of seeing things".

– through the joint development of appropriate governance, we gain energy for action and greater impact to build synergies on the selected projects, while leaving enough autonomy and initiatives to the different partners of the local system (section 3.5).

3.3. Building a project or a group of compatible projects within the territory

3.3.1. *Projects as a means for successfully developing tourism systems*

To develop the attractiveness of their territory and the performance of their activities, the members of a tourism system must constantly instigate and renew collective projects. Destination contracts, for example, are construed as projects, which mobilize or enrich the available resources with the goal of drawing in tourists.

Each project is based on dynamic modeling by the actors of the tourism system in question, or at least by the group which becomes its driving force. "The nature of an organization is what people put into it when they built it". It therefore depends on "what underlies and justifies this construction, regardless of the concrete organizational form […] in which it is embodied" (Bréchet and Desreumaux 2018, p. 21).

For Le Moigne (1990, p. 170), every system denotes "a set of teleological actions[15] within an active environment". It is a network of operations, information and decisions. Its attainment is intelligent and has memory. It has memories linked to its history and the decisions that have already been made, even though these were made by different actors.

Depending on the projects giving life to it, a tourism system may change in nature and operation. For example, the Amnéville site was no longer the same after Doctor Kiffer's death, nor was the Louvre-Lens, following the political changes which affected the region/department where it was located. But from the point of view of the territory and the strategic continuities or discontinuities it experienced or underwent, a memory certainly remains.

3.3.2. *How to initiate and bring a project to life?*

Any tourism system may be analyzed as a "project system" (or a constellation of projects that are partly compatible, partly tensional and partly capable of being

15 Guided by goals (implicit, stated flexibly or formalized in strategies).

organized optimally on the basis of a common project). According to Boudinet (1996), a project must find a balance between its efficiency, which requires structuring and overall framing, and its mobilization potential, which requires lasting and consistent exchanges between stakeholders.

To launch a project, it is necessary to agree on a problem situation, the actors to be engaged, its aims or goals and to coordinate the implementation of proposals. This revitalization has expected and unexpected effects, which may lead to new thinking.

Within the frame of a territory, environment and local issues lead to various findings: observable realities, difficulties or constraints, opportunities to be seized or set up, availabilities which can be exploited. These dimensions have been formalized into strategies by tools such as SWOT[16], or SVOR. Among the constraints, those which are difficult to tolerate prompt changes that are facilitated by opportunities or potential flexibilities in the existing situation. We must also accept insurmountable constraints and, among the availabilities or opportunities, choose those thanks to which we may set up a project that is attractive, realistic and acceptable to the various stakeholders.

Any project requires an intellectual construction of a collective nature, based on realistic data, feasibility of implementation and consultations, capable of engaging as many actors as possible in the system in question.

From the project's perspective (Boudinet 1996), a system's boundaries are mobile and have variable geometry. They depend on the actors who take part in collective interactions, accept certain common rules or engage in learning processes. But what is true for the territory is also true for the companies coexisting there, and which themselves do not always have clear boundaries between what we would judge to be their "interior" and their environment.

Instead, we know porosity exists and stakeholders extend beyond their limits, starting with the system's customers, the partners and all of those who are indirectly interested, without considering themselves as integrated members. This is even more the case for systems made up of networks of individuals, businesses and communities, independently of the institutional barriers brought by territorial limits (from a geographical, touristic or political point of view).

16 Matrix conceptualized in the 1960s by the Stanford Research Institute (SWOT stands for Strengths, Weaknesses, Opportunities, Threats, whereas SVOR stands for Strengths, Vulnerabilities, Opportunities, Risks).

For Bréchet and Desreumaux (2018, p. 182), any project-based business is "entrepreneurial and multidimensional in nature". To reach or respond to its goals, each project strives to integrate the necessary groups, with the problems and perspectives that this generates. Surely, it is also "a construct of materials, schemes, rules and skills". But, at the same time, it is the constitution of a collective struggling with evolving circumstances. And this construction is the "collective's common good".

As this implies being committed to cooperation, the participants must be able to anticipate benefits or advantages that make sense (from a perspective that is not only economic but also integrates all of the richness of human behavior).

In a certain number of cases, projects are driven by key actors with an entrepreneurial profile, capable of implementing relatively innovative and partly risky strategies (Nobile 2021).

When it comes to territories whose challenges are both economic and social, the dynamism of elected entrepreneurs may help build projects designed as vectors for global development (Amnéville, Louvre-Lens, Nantes and, without the same success, Cognac, Dordogne or Morbihan). But the more complex (or even vague or equivocal) the information to be collected is (Martinet 2016), the more all local actors need to be mobilized, their conflicts solved and realistic governance found in terms of cooperation. This will be explained in sections 3.4 and 3.5.

3.3.3. *Application to tourism systems*

In the case of a tourism project, experience and the literature have focused on four complementary points to be exploited, which correspond to the idea of an attractive destination. It is in this direction that the projects (or clusters of projects) that will be implemented within the territories should be organized:

– First of all, the portfolio of natural or created resources must be highlighted as comparative advantage. This concerns assets relating to tourist spaces and landscapes, as well as to the historical, artistic, cultural and natural heritage. The more unique, authentic and difficult to reproduce the character of the main resources is, the more original its attractiveness will be. Thus, it can be exploited for thematic specialization[17]. We can even have fun with certain stereotypes and stage them with a hint of humor[18].

– Projects must then rely on competent people and an appropriate workforce because the quality of practices is essential for the proper exploitation of resources. At the same time, we must have the infrastructure or capital required for their maintenance and development.

– On the financial level, the tourism industry thus constituted must be competitive, through attractive prices, as well as likely to generate profitability. This involves obtaining the dynamic adaptation of supply to demand. This is where, in the projects implemented, the search for overall efficiency can intervene, with the idea of facilitating innovations and coordinated (public and private) efforts, in order to guarantee the smooth development of the tourist supply. The cost/quality ratio must remain satisfactory over the whole thus recomposed.

– Finally, the destination must be able to summon institutional support (and in any case, means of communication) to strengthen its tourist image.

Tourism development policies involve challenges of three types (with differential priorities, depending on the sites and their current situation):

– the promotion of the territory through the development of new resources;

– the structuring engineering (overall organization, optimization of resources);

– the networking of actors to enhance their collective effectiveness.

We may characterize those challenges in a different way by emphasizing two points which usually play a major role in project design: the search for attractiveness and the constraint of efficiency.

3.3.3.1. *Developing attractiveness*

Attractiveness involves building (or skillfully using) differential visibility. What can the destination offer us to makes us want to go there, and prefer it to other stays or travel opportunities?

---

17 Mountain, coastal, urban tourism, etc. Or family, cultural, taste, health (thermal treatments), sports, etc. Or polyvalent, with a strong local identity.

18 Refer to the work by Guérout and Mauduit (eds) *Histoire des préjugés*, Les Arènes, Paris, 2023. All countries, along with their tourists, are targeted. We could mention the recurring discourse on "arrogant and ill-mannered French people", pp. 229–235, or "unpleasant waiters", "sully taxi drivers" or the dirt of the country (*ibid.*), to set up communication backfires.

We must use our imaginations and stimulate the desire to visit, within a framework that must remain realistic.

For this, the challenge is to use a range of complementary means so that the destination is recognized and appreciated from afar as somewhere pleasant and feasible (accessibility, conviviality, multiple opportunities).

This amounts to the products offered (two potential lists may be found in Boxes 3.1 and 3.2[19]) and how to make them known (which requires mastery of promotional communication and digital tools).

The purpose is not to offer everything, but to organize the supply in such an original way that it not only meets essential needs but is also presented in an attractive and skillfully publicized form.

Getting away from it all: finding ourselves somewhere else, experiencing a "change of scenery", recharging batteries, free, relaxing and possibly doing something different or indulging in pleasurable activities that are difficult to practice in normal times.

The fascination with monuments, sites, experiences, events which are of interest in themselves and which can be explored in depth (from cultural heritage to natural spaces and sites).

The need to experiment, including risks, unforeseen events, or with the obligation to chart our own path within a propositional setting that may remain vague, fuzzy, dense, etc.

The desire to rest, to relax (to the point of *farniente*).

The surprise effect (being amazed and impressing your loved ones with what you have seen or discovered).

The influence game (using travel to dominate, influence ("being an influencer"), advise, play the role of a *leader*).

Culture-hungry: learning, developing knowledge, sharpening our curiosity about people, situations, cultural or artistic experiences.

19 We will refrain from calling into question the main categories of supply proposed by geographers (Chapter 2): rest, discovery, recreation, sociability, shopping, but will present a more detailed version which can correspond to further differentiation (and groupings on sites), with the aim of combining the plural local supply with approaches producing competitive advantages.

The taste for encounters (convivial, astonishing, friendly, romantic, fortuitous, calculated, repeated, etc.).

Sensory pleasure (comfortable rooms, good meals, cellar visits, etc.).

Sports practices (natural practices exploiting places ad hoc, ranging from mountaineering to fishing), technical practices (practicing what we love doing, regardless of the location).

Wanderlust, the "tour de France" or cruise atmosphere optimizing stops and visits (at a running pace).

Obsessions, hidden desires, pretexts: dark tourism[20], sexuality, substances, etc.

Box 3.1. *Supply options related to possible expectations*

Traditional itinerant tourism.

Fully or partially packaged traditional tourism.

Free tourism starting from a fixed location.

Nomadism (traveling by motorhome, van or converted truck).

Organization of unusual adventures. Distant tourism (change of scenery).

Local tourism (return to basics, visits to family, reduction of budgets or journeys).

Tourism at home (website consumption without going to real places or taking time to investigate before choosing).

Fusion trip (alternations or mixtures between the city and countryside, rest and exploration, etc.).

Slow tourism (taking time to disconnect from ordinary reality).

Regenerative tourism, conducive to health (or guaranteeing health safety).

20 Dark tourism involves visiting places marked by natural disasters (earthquakes, explosions, accidents, fires) or humanitarian tragedies (concentration camps, sites of massacres or cruel activities). The reasons for this are multiple, in a similar manner to the attraction felt towards reporting on dark news or, in imagination, towards dystopias or crime series (as seen in film productions).

Culinary or enotourism. Activity tourism (cultural, sporting, hiking, etc.). Creative tourism (to learn an art or a craft).

Family tourism (oriented by the needs of children).

Ecotourism (contributing to the protection of natural spaces, exchanges with local residents).

Commitment and responsibility tourism (activities at the service of people, helping to restore heritage, etc.).

Collective accommodation tourism (to save on costs and meeting unexpected people).

Tourism related to events we participate in.

Mixture of tourism and work ("workation" as in work+vacation, "bleisure" as in business+leisure).

Last minute tourism (seizing opportunities at reduced budgets).

Box 3.2. *Supply configuration in response to trends*[21]

3.3.3.2. *Increasing efficiency*

In economics, efficiency always results from a favorable relationship between what we spend and what we recover through the benefits of actions or investments.

In a tourism system, collective effectiveness is both global and dependent on the results obtained by all of the individual actors making part of it, even though their expectations and needs are not the same. In practical terms, the network of actors present is made up of a multitude of production units, each with its own business model.

None of this should be overlooked, and even less so, dismissed.

We should bear in mind that collective results partly come from independent or autonomous actors, and that it is necessary to manage the balance between cooperation and competition as best as possible[22]. In any Localized Tourism System, it is thanks to great managerial flexibility that we can regulate coopetition.

21 For this picture, we have compiled around 20 websites (from public organizations and associations representing the actors of tourism, influencer blogs reporting on fashion, etc.).
22 This point is addressed in section 3.4.

3.3.4. *The characterization of ongoing projects*

If we now look back at the cases studied in Chapter 1, we may immediately note that projects require clarity ("easily understandable by local actors"), operationality ("they lead to concrete challenges around which collective action may be organized"), dynamism ("they rely on qualitatively and quantitatively adapted forces") as well as continuity ("they require time, to be activated in the long term"). Every time one of these characteristics is deficient, it ultimately weakens or disrupts the project.

In general, we find clarified projects: "spreading a product which has already proven itself", Center Parcs; "abolishing unnecessary territorial boundaries", Léman, Dordogne; "emerging from an economic crisis thanks to tourism", *Autour du Louvre-Lens*, *Voyage à Nantes*, Amnéville; "innovating in tourism", TIL; "injecting tourist energy to a sluggish territory or one losing impact", Saint-Jean-de-Monts, *Saumurois*, Morbihan, Cognac. Sometimes, the projects are mainly defensive or involve local adaptation (Destination Anjou, Mélusine).

Even when projects are clear (at least according to their initial intention), they are not always operationalized. Sometimes it is more of a show-off or mimicry than an asserted desire accompanied by a seriously constructed action plan, as in Saumur; or else, the translation of the issues into forms of action has not been satisfactory, as for Cognac, the TIL or the post-startup phase of the Louvre-Lens.

In too many cases, projects partly lacked energy or support. The projects that worked best (Nantes, Saint-Jean-de-Monts, Amnéville at the beginning) benefited from strong political momentum and local mobilization. It is also at this level that we appreciate the importance, for projects, of dynamic governance and networks supporting collective action.

Finally, many projects suffered from discontinuities or delays in their implementation due to institutional constraints (Saint-Jean-de-Monts, Morbihan), conflicts between the actors (Dordogne) or the loss of key players (Amnéville).

3.4. The revitalization of the networks of actors

This section studies in detail the driving forces of a local system, namely its members, their types of involvement, their levels of competence and how they interact with one another.

The territory offers geographical proximity advantages to the actors who work there and contribute to a supply related to the site's charm and to the organization of programs making tourists feel welcome. But physical proximity is nothing without the collective intelligence of actors to optimize their functioning, beyond competitive logic and particular interests.

This is why we must deal with another type of proximity, this time relational, which lets us get along, cooperate and build concerted projects. This type of proximity involves exchanges and ties and depends on the quality of the networks of actors (communications, cooperation, rounding-up of activities).

To turn the territory into a collective asset, we must jointly act on its access, the complementarity of its attractions and its general coherence, so that the different clienteles can share this space without tension. Internal or external circulation (e.g. on opportunities for neighborhood) should be facilitated.

In hierarchical organizations (companies, local authorities), exchanges are formalized by statutes, and relationships regulated. But this normativity slows down relational opportunities and blocks initiatives which could be useful for collective action. We can avoid this drawback thanks to a fair proportion of work and cooperation in networks, through autonomous regulation in the sense of Reynaud (1994), even when the activities are constrained by management or procedures (formal regulation).

When systems are freer and actors (independents or managers) are not inhibited by hierarchy, the quality of their networks makes them more efficient in the achievement of cooperative projects. From the moment these networks acquire the capacity to inform their members, train them or help them to produce, the tourist territory takes on more creative and operational substance.

Following this, we will analyze networks based on their general properties (section 3.4.1) as well as the types of ties that are formed within them (section 3.4.2). Let us recall that cooperative and competitive operations coexist.

Then we will see how active and dynamic networks are regarding what we call innovative environments (section 3.4.3), capable of mobilizing the action of their members and making them proactive in relation to changes in the environment.

We will then address the characteristics of the two main categories of actors (section 3.4.4):

– public agents and private employees who depend on organizations present on the site;

– the leaders of these organizations and independent actors who have freer interactions.

To make networks more effective, we must increase the interest that local actors find in them. This encompasses strengthening intermediation tools, whose possible modalities we will describe in connection with tourist activities (section 3.4.5).

Finally, regarding the systems explored (the cases in Chapter 1), we will evaluate the status of current networks according to the above-mentioned dimensions (section 3.4.6).

3.4.1. *Network characteristics*

Networks are found in almost all complex systems, which demonstrates their relevance and the numerous uses that can be made of them (Barthélémy 2022[23]).

Their structure always involves propagation paths (in energy or information, for example). They contain nodes of interactions which may act as a crossroads for exchanges, as places of storage or action, bifurcations, interferences, etc.

From these networks, relational maps are drawn with opportunistic or obligatory types of circulation, places for communication or production, stocks of information or knowledge.

In social psychology, the groups or communities of actors are always defined according to the relationships they develop. A community is a system where the relational links are considerably more numerous between its members than with the outside, even though intersecting marginals will always exist (i.e. people who are on the edges of their group, but who participate in other groups at the same time).

In systems with economic issues, such as Localized Tourism Systems, networks are nourished by local work experiences between productive actors, their customers and various partners. They develop strategic or tactical opportunities among themselves, which may lead to synergy. However, when networks are poorly organized, they could provoke conflicting or deficient relationships, leading to inertia or resistance.

We can always enrich existing networks with inputs, for example by integrating external partners or newcomers who arrive with their connections and make them

23 Barthélémy, M. (2022). *Le Monde des réseaux*, Odile Jacob, Paris. This author, originally a physicist, proved that "network science" exploded over the past 20 years. He showed the extent to which networks are universal answers for building complex arrangements in physics, engineering, in living matter (blood or neuronal networks, for example), or in social interactions. In psycho-sociology, Moreno developed sociograms, or active interpersonal networks, as early as 1934.

accessible to others under certain conditions. They may also be improved by creating channels for intermediation (see section 3.4.4).

The analysis of networks allows us to see the state of relationships between people or groups in a Localized Tourism System. It involves diagnosing the social ties found there.

3.4.2. *The nature of ties within networks: strong or weak, cooperative or competitive*

3.4.2.1. *Cooperative modes*

To activate Localized Tourism Systems, we have to gather as many cooperative forces as possible within existing networks.

But obviously, as in the sociograms by Moreno (1970), we can find positive affinities, situations of relative withdrawal, and even hostile or conflictual relationships between people. The more agreements and collaborations there are, the more a form of "community" is born between the actors, whereas in the opposite case, the tourism system is no more than a mere political discourse or a virtual project for land developers.

In an operational context comprising territorial cooperation, we may play with a variety of ties (strong and weak). This is what Granovetter (2000) taught by emphasizing the relative importance of weak ties. Normally, the strength of ties is characterized by the combination of time spent together, emotional intensity and intimacy between two people. In this sense, strong ties are those we have with family, close friends, colleagues or partners, with whom we have sustained and frequent relationships. While they have important advantages, they also require time dedicated to maintenance and regulation. Weak ties are simple acquaintances, but at their level, they also have certain strength, because, when diversified, they enable the penetration of networks other than those made of strong ties; they offer good potential for opportunities. On the contrary, there is no tie if there are no interactions (think, for example, of two restaurateurs who know that the other exists, but who never speak to each other).

In networks, it is in our interest to create positive ties by reinforcing the strong ones and developing recurring or occasional weak ties.

3.4.2.2. *Conflicts and coopetition*

We must also accept that not all relationships are positive, and take into account the tensions, disagreements and sometimes substantial conflicts between certain actors.

Among the sources of tension, there are the types of actors who have a different background and ideology. Even though they are compelled to act together in tourist systems, they sometimes hold heterogeneous representations. In the current logics seeking to reconcile public action with private practices, or to combine cultural contributions with tourist requirements on the territory, we may still find divergences in goals, mission conception and the concrete functioning between the public and private sectors, and between the worlds of tourism and culture.

Finally, without neglecting individual hostilities due to sometimes serious relational liabilities, we must acknowledge the undeniable phenomenon of coopetition.

Conceptualized by Brandenburger and Nalebuff (1995), "cooperative competition" (or coopetition) is drawn from game theory. It describes situations where it is better for competitors to work together rather than come into conflict with each other. As soon as complex interactions emerge between economic or political actors, we may observe back and forth oscillations between competition and cooperation. Even fierce competitors may help each other on aspects regarding collective defense or the general organization of their shared space of action. And on certain occasions, they need one another.

In Localized Tourism Systems, all actors have an interest in helping each other to promote their territory, consolidate it, enrich it or defend it against external "enemies". But that does not hinder them from contending in the areas where they are competitive. Balance needs to be found between collaborative challenges and keeping emulation fueled by partially divergent interests.

The types of conflicts are often linked to the actors' characteristics:

– Politicians fight for the power of collective direction and for decision-making positions: who will decide in the last resort, who will impose their law, who will access the best place?

– Experts seek to stand out for their prospective vision or technical skills: who is ultimately right about what needs what to be done, and how?

– Field operators wish to continue to develop (or survive): how will I preserve my interests in the face of cooperation?

These differences of place or interests may be counterbalanced by collective trust and the analysis of opportunities inviting cooperation. Power is not a zero-sum resource; we can increase it globally (by acquiring more impact on the outside) and redistribute it with greater amplitude.

As much as possible, we must try to get out of personal conflicts and the battle of the egos, because these conflicts confuse interpretations, limit collective lucidity and expend considerable energy which could have been better spent elsewhere.

3.4.3. *The impact of innovative territories*

The importance of positive interactions is found in two very similar themes: "creative territories" and "innovative environments".

3.4.3.1. *Creative territories*

Through their active organizations and networks (Glon and Pecqueur 2016), territories may generate varied initiatives beyond the strictly economic aspect, by bringing out collective intentions. Local creativity rarely stems from a standardized injunction, but rather from "effervescence".

Territorial proximity requires both "spatial" and "social" proximity. There is a need to build "sociability" (relationships, exchanges of ideas or projects), while creating "a collective dynamic of appropriation and valorization" in relation to the assets and potential which emerge from interactions (*ibid.*, p. 11).

"Thanks to the local network", the territory can "assert its specificities" in a proactive manner, based on resources which may be historical or coming from new impulses. The latter help the territory to enrich itself and build numerous positive effects around innovative contributions or ex nihilo implantations (e.g. JIM, Jazz in Marciac, created in 1978, has become a global phenomenon with tourist, economic and even real estate consequences). Another example: cities may profit from an event ("the capital of culture") to transform themselves (Mons[24], in Belgium, exploited a short sojourn by Van Gogh in the Borinage and stated that the area allegedly "sparked his artistic vocation", whereas this is no more than a resource anecdote).

Creative territories combine the commercial and the noncommercial, social initiatives and economic contributions. They do not necessarily acquiesce to the institutional limitations of the administrative territories where they are set. Glon and Pecqueur (2016) spotted the subjective and multidimensional aspect of the resources used:

24 In virtue of his extreme popularity, Van Gogh also attracts audiences at Haute Veluwe Park (the Netherlands) and Arles (France). There is even a (much visited) Christ tomb in Japan, and a site called the Twelve Apostles in Australia.

– they are material (a company, a museum, a forest, a landscape, everything that fits into concrete spaces);

– as well as combine with the intangible (symbols, reputation, information or knowledge, ecological, cultural or social attractiveness, human skills and experience, capacities for action, taking the initiative).

3.4.3.2. *Innovative environments*

"Innovative environments" refer to productive actors located in a territory (companies, institutions, independents, political decision-makers, various collectives). There, they develop multilateral interactions, within an environment rich in emerging potential (Tanguy and Uzunidis 2016). The quality of the ties established by these actors makes it possible to take advantage of the region's resources and contribute to the creation of activities or jobs. They deploy partnerships and stimulate each other through dynamic confrontations.

These tangled ties relate directly to what we have described as active networks. They amplify exchanges, contacts or cooperation, if necessary, relying on operational systems which constitute the nodes.

At the level of economic actors (managers, entrepreneurs, creators, local facilitators), an innovative environment and its networks stimulate the arrival of financial resources, the development of relational capital and useful knowledge, as well as the establishment of infrastructure and support services. Local exchanges are more confident and foster effective collaboration, even in situations of relative competition.

In simpler tourist sites grouped in small areas, networks need not be so complex.

In tourist systems spreading over larger territories, the emergence of networks may be multiple, such as *Autour du Louvre-Lens*. They correspond to complementary developments which may interact, overlap or create redundancy (but redundancy is not without interest for mutual knowledge and learning). The fact of there being competing (but harmless) networks should be interpreted as a sign of vitality.

On the contrary, certain territories may be described as "inhibiting environments" if they suffer from:

– an insufficiency of concrete resources mobilized towards their fields of development, be them in the form of tourist opportunities, financial resources or infrastructure supporting tourist reception, accommodation and circulation;

– a lack of local actors capable of getting involved in the necessary changes, of adapting to them or being innovative. This may be due to surrounding nervousness, a skills deficit, inappropriate networks or not finding the desired tone;

– excessive conflicts between a certain number of key actors, something which blocks collective energy and paralyzes project implementation. These recurring tensions may also disrupt emerging interactions and reduce trustworthiness.

Action is easier in a territory having already experienced the accumulation of knowledge, the aptitude for learning and experimented with productive creativity. In short, things evolve more quickly the greater the territorial capital.

As Tanguy and Uzunidis (2016) observed, today's economy heavily relies on knowledge. When assessing the value of a good, what counts is not so much its manufacturing conditions, but its design qualities (upstream) and the know-how put into selling or distributing it (downstream). Therefore, two types of processes appear as decisive:

– the development of human capital and intangible resources. This includes the desire to innovate, the ability to acquire training and to accumulate useful knowledge;

– the advantages that local actors find in networking and building collective emulation or organizational cooperation.

Tourism needs to progress in this knowledge economy. Having creative resources at our disposal depends on building them beforehand. Not through improbable transplants, but through all kinds of training sites, interactions and collective learning.

3.4.4. *The actors to be mobilized depending on their status and framework of action*

Within the networks, we often find front-line actors (business managers, craftspeople and retailers, political leaders, territorial managers, consultants, independent operators, etc.). But, behind them, in private and public organizations, there are agents and employees whose commitment and contribution are decisive for collective activity. Moreover, some of them take part in networks because they intervene directly in events, they have expertise at their disposal or they coordinate operational teams.

3.4.4.1. *Reducing hierarchies for the benefit of networks*

Even in organizations (communities, businesses), we should combine hierarchical relationships with networking[25]. Regardless of where they are, people will not mobilize unless they are recognized in collective action and allowed to take initiatives or responsibilities. Networks facilitate the logic of subsidiarity (delegating the power to act to those who are closest to where the action takes place).

A previous work (Louart 1980) showed how regular exchanges between craftspeople on questions regarding their activity – simply by sharing ideas and joint stimulation – enabled them to increase their turnover, rejuvenate their products and innovate commercially. Two years of discussions, one evening every fortnight, had given them the desire to renovate, to imagine and to undertake projects. This, in turn, made their businesses grow and increased profitability.

As for companies, Landier (1991) relied on numerous studies and expressed the following:

– "The company of the future will necessarily be an intelligent one. It will have to mobilize ever more brainpower, accept pluralism and encourage the diversity of individual initiatives (the only way to respond collectively to unforeseen events). It will base its success on the personal development of each of its members" (*ibid.*, pp. 17–37).

– "The relations between companies (including competitors) cannot be reduced to economic warfare and confrontational relations. Establishing cooperation is a condition for success [...], made possible by a game of subtle diplomacy" (*ibid.*, p. 200).

– Through active networks, in companies, "one shifts from a morality of subjection to a morality of freely agreed cooperation". Thanks to "a common vision [...] which transcends particular interests [...], relations of domination must give way to relations respecting the freedom and intelligence of others" (*ibid.*, pp. 201–202).

3.4.4.2. *Giving meaning, pleasure and dignity to work*

Beyond networking, many employees in tourism companies need to be offered attractive, better-paid employment, with decent working conditions. While cultural workers benefit from a specific and valorized unemployment allowance, tourism

25 According to Pagels (1990, p. 128): "Most real systems (in nature and living things) are a mixture of hierarchy and network". For Reynaud (1994), social systems always combine formal regulation (e.g. procedures) with autonomous regulation (provided by each of the actors present, according to emerging needs).

workers get by as best they can. Relative precariousness offers them nothing more than ordinary unemployment compensations or the current logic of temporary work.

Until recently, this system had been justified by the recurring problems of small business owners (themselves without any guarantees in their activity), the paternalistic habits in craftsmanship and retail, and the opportunistic abuse of socially unaware workers. For their part, large structures have not always bothered with social ethics, for example, by using dependent immigrant workforce for their benefit.

Some still consider tourism as a fluctuating activity, subject to customers who are not easy to please, and where it is also difficult to take care of employees. However, we should think the other way around, as in Nayar's (2010) theory on the symmetry of attention. For this author, it is by taking excellent care of the company's employees so that they will feel encouraged to take good care of customers. By respecting them, they will feel encouraged to respect tourists and show ingeniousness.

After the disruptions provoked by the Covid-19 pandemic (2020–2022), the relative mismanagement of tourism staff (there are happy exceptions, though) led to a phenomenon called "the great resignation" (or the "big quit"). Many employees left the sector, which is continually struggling to find new staff. This state of affairs is amplified due to three complementary reasons:

1) Today, even small employees are mobile (especially the youngest), with a taste for international jobs associated with travel. They compare their status or remuneration from country to country, disadvantaging the French context.

2) Universities have developed the training of tourism executives, who often complain about being underestimated in relation to their skills or bound in the initiatives they are allowed to explore.

3) Social networks, on the Internet, also lead to social comparison. These are propagation vectors when employment situations pose a problem. When there are no dynamic networks in concrete tourism systems, we look for them elsewhere to serve as bases for evaluation (Facebook, Instagram, TikTok, etc.).

Localized Tourism Systems could improve the Human Resources Management of local employees by serving as delegated operators for small businesses overwhelmed by their daily work. There are at least three types of possibilities:

– Employer groups for core staff. There are specialized groups (the companies' collective "GEIC", "200 sur le territoire") that train candidates without skills, as

McDonald's does with youngsters in working-class suburbs, who are not sufficiently integrated into society and are hired by the franchisees.

– Jobs managed jointly by several companies, for example, to accommodate executives who do not work full time on a single activity.

– Territorial Human Resources Management through job security policies in a given geographical area. Employees having to leave a company due to economic reasons (closure, drop-in activity) would be considered priorities in terms of reemployment, so that they can keep their home and their personal networks.

We are yet to vitalize collective action and foster the desire to act, create or progress together. During the Covid-19 pandemic, some businesses and public administrations used "recipes" that worked well. They found solutions to many of the problems suffered because of the crisis. However, these were not new. Such practices had been available for a long time[26], but bureaucratic routines or hierarchical habits had dismissed them: giving people back the power to act with more autonomy, recognizing and spreading good initiatives in the field, relying on the dynamics of communities, daring to experiment with new possibilities and learning from them.

3.4.5. *The importance of intermediation*

As we have already mentioned, networks need to be constantly stimulated and energized to maintain their interactive strength. If they are launched without the means for *reenergizing* them, they run out of steam and lose consistency, projective capacity or creativity. This is observed in certain tourist areas that have benefited from start-up or relaunch impulses but have failed to set up dynamic maintenance in the medium and long term.

In hierarchical systems, it is the political power (public authority) or institutional actors (leaders) that order activities by imposing forms of organization and task distribution.

In networks, intermediation schemes play this role, facilitating exchanges, creating favorable spaces and hosting meetings where the members of these networks will find what they need to:

– give hope, courage and collective confidence to one another;

26 In the 1970s–1990s, throughout Europe, corporate democracy developed: autonomous production groups, socio-technical approaches, companies with flat structures, co-management, self-management. The recent concept of "liberated companies" has only brought some of these experiments up to date.

– receive or find resources and support;

– accept their relative competition, sort out their disputes (through arbitration) and take a leap forward, thanks to common issues and territorial projects;

– acquire skills (training, knowledge sharing) and build new tourism products together (production and innovation).

These intermediation schemes can be translated into various concrete forms. These include the following:

– Materialized structures (offices, platforms, hubs, centers of activity, formal groupings, contracts which imply mutual commitment, such as destination contracts). In current practices, the exponential development of all kinds of third places, enabling co-working or the installation of project laboratories.

– Relational processes (meetings, conferences, demonstrations, events). Events are not just a means for luring tourists or mobilizing the inhabitants of a region, but also a sort of melting pot for meetings, initiatives and learning (Piriou et al. 2017). In around half of the cases studied (most notably, the *Voyage à Nantes* and *Autour du Louvre-Lens*), the production of events contributed directly to developing tourist activities or establishing their continuity.

– The gradual construction of a common language for effectively communicating with one another, enhancing understanding and clearly discussing the projects to be implemented. This common language is fundamental for three reasons found in strategic or commercial literature: 1) sensemaking (having a shared interpretation of reality, especially of what we may call the territory's resources); 2) storytelling (inventing stories, producing narratives or accounts, so as to exploit them in marketing and promotional brochures); 3) issue selling (skillfully selling the tourist destination by highlighting what directly or indirectly makes it appealing).

3.4.6. *Network evaluation within the Localized Tourism Systems studied in this work*

In the cases studied, it was difficult to accurately measure the reality of active networks, except when the analysis directly addressed horizontal-type networks (*Destination Anjou*, Mélusine) or in connection with a private organization having a dominant hierarchical nature (Center Parcs).

Within the territories, networks diversify even further as their size increases (Louvre-Lens, Cognac, Morbihan). The lack of intermediation or engaging governance clearly penalizes them, as do conflicts between the actors and the lack of enticing challenges (projects making people want to get involved).

We often neglect the revitalizing effects of the participatory management of human resources (both in public and private companies). This theme has been insufficiently developed, except to complain about workforce shortages (in the private sector), or an administration lacking flexibility (in the public sector). But behind this observation, there are teams to be identified and trained.

However, in isolated cases, we perceive the interest in stimulating the local environment and inviting groups to interact (to overcome the profession's compartmentalization). We also value the benefits of opening up to the outside world and enriching the tourist potential by breeding useful actors, extending itineraries or multiplying visit opportunities. Such an opportunity was evident in Nantes, the Vendée coast, Lake Geneva, *Autour du Louvre-Lens* and even Center Parcs. Unfortunately, it did not fully succeed in Morbihan, Dordogne or the Cognac region (undeniably between the wine cluster and the actors with tourist priorities, who were far more dispersed).

Last, but not least, the forms of intermediation proposed primarily assumed an institutional character: the creation of associations, of semi-public and public companies; the implementation of tools supposed to facilitate the creation of a cluster; destination contracts. The more informal or individualized tools, which make it possible to enrich fertile proximities, have not been sufficiently explored. For example, the installation of third places with plural goals, or the multifaceted use of alluring events, etc.

3.5. The establishment of appropriate governance[27]

In the spirit we have approached them, networks are only of interest provided that they contribute to build and legitimize collective action by the actors involved. In order to achieve this, it is necessary to adopt holistic governance (or a network of partial and more or less compatible modes of governance).

In this section, we will analyze governance and the forms it may take. First of all, we must ensure consistency between the action goals and the game rules to attain them. These goals and rules must be acceptable to stakeholders in the tourism system, which brings us back to the definition of legitimacy (section 3.5.1).

---

27 The following presentation is inspired in managerial philosophy. It still needs to be translated into legal rules imposing governance systems, particularly in the public sector. Today, the composition a priori of certain Agglomeration Communities (AC) or schemes is not favorable to their democratization. We have recently seen the consequences of this in the governance of French sport (football, rugby, tennis, etc.).

Technically, there are different ways of implementing governance (ranging from paternalism to bureaucracy). These multiple approaches correspond to well-known models (priority given to shareholders, public supervision, integration of stakeholders, autonomous management of the common good). Most of these models exist in all or part of the tourist configurations. They more or less directly condition what we observe in local systems (section 3.5.2).

Given the heterogeneity of actors, the fluctuating nature of tourism systems and the need to activate networks, the most appropriate forms of governance are collaborative or mediated (section 3.5.3). Mediation should be used when the goal is to reduce tensions within the system, before opening up any voluntary spaces for cooperation.

Collaborative governance requires largely nonhierarchical forms of exchange (section 3.5.4), with as few narcissistic or authoritarian leaders as possible. And they must be realistic in terms of findings and plausible futures, acknowledging the contexts of tourism development.

This type of governance is inspired (section 3.5.5) in the works of Ancelin-Schützenberger (1971) and Saint-Arnaud (2009), which show the extent to which the collective energy requires agreements on goals and cooperative solidarity. At the same time, we must reduce conflicting tensions and the desire to escape of those who are not fully committed to the system.

As in previous sections, we will conclude with a brief evaluation of the governance characteristics in the tourism systems studied (section 3.5.6).

3.5.1. *How to build legitimacy?*

We cannot develop a collective strategy without legitimizing its bases, actors and goals. In an article which summarizes his perspective, Reynaud (1994, pp. 204–210) explained how legitimacy is developed. He detailed its four major ingredients:

– "Every rational actor arranges its usefulness by adopting satisfactory solutions to the problems at hand. Even when the relationships to other actors are power relationships […], the rational actor can contribute to the system's balance and regulation".

– "The interests of the actors are not natural data […], but the by-product of having developed an action system. The rules of the game […] are a social construct which is only stable to the extent that the actors grant it legitimacy".

– "To say that the rules of the game have legitimacy is not to say that they are inviolable or definitive [...], but that their violation has a price and they cannot be changed by unilateral decision". Of course we can cheat or defect, but this often entails a high cost (loss of essential links, damage to reputation, exclusion).

– Legitimacy is built in two possible (and often complementary) ways: via the results obtained in previous actions (reality principle) or through the exchange of mutual recognition (solidarity principle). According to Reynaud: "it is the legitimized collective actions that create the collective actor". Even though legitimacy is based on the positions of authority of key actors, or on rationalized management systems, legal elements, etc., it cannot be founded from the outside. In essence, it is the result of collective actions.

From these general principles are derived the role of networks (to carry out actions or build conviction), and the need for governance (to support decisions concerning collectively constructed legitimacy).

3.5.2. *The main types of governance*

The term governance includes the fact that, even in hierarchical or fixed systems, the actors and their environments are changeable[28]. An authoritarian actor may give way to a constellation of micro-decision-makers who do not always agree with one another; a motionless and favorable context may transform into a suddenly disrupted situation. At this point, it may be suitable to enrich the reflection with the "regulation" paradigm, in the manner of Crozier and Friedberg (1977) or Reynaud (1989), for whom social reality is a regulation exercise, not the consequence of rules fixed once and for all. We must take into account the manner in which rules are created, transformed or disappear, through never-ending collective negotiation.

In some cases, multiple systems coexist (with ongoing tensions or shifts between systems). We may then ask two questions:

– How is it possible to articulate sub-groups, each with their own governance, but still functioning within a tourism system whose coopetition forms must be complementary and not exceedingly harm one another?

– What is their evolutionary potential to optimize operation and (economic, social, political) results? Should we privilege the form of governance deemed as the most appropriate? These are territorial challenges, not merely concerning economic viability.

---

28 Governance is comparable to a mobile government that constantly adapts to fluctuations in reality.

Let us briefly recall the history of these models.

Traditionally, the forms of governance have been paternalistic or familiar. Many are still existing, particularly in tourism Very Small Enterprises, where managers are simultaneously business owners, and even in larger groups, where historical leaders have been imperious coordinators (as in Amnéville in the past, or at Puy du Fou still to this day.

There is shareholder governance when activities depend on private consortiums (e.g. hotel or casino chains, Center Parcs, leisure parks or zoos multiplying their implantations).

And there is public governance when it is the representatives of the state or the communities who manage activities, based on public law administrative provisions. In France, politicians tend to rule as soon as they gain a legally attributed "competence[29]".

These forms of governance have their limitations, nonetheless. Some researchers and managers have considered a partnership-based governance, described for example in a collective article by Bréchet et al. (2015). The authors abound on the theory of stakeholders and attribute speaking or arbitration rights to actors other than shareholders, elected officials or managers. But this theory does not fundamentally call into question ancient practices. For its author, R. Edward Freeman, it is only a form of expansion, inviting the reflection of other types of influential actors apart from shareholders (customers, employees and certain strategic partners), without questioning profit maximization (Bréchet and Desreumaux 2018).

The economy of conventions is more open since it considers organizations as a way of collectively adapting to the limitations of individuals. It adds ethical, political and social dimensions to the economic imperative by insisting on the need to create institutions bearing significance, communicating rules and values, beyond the logic of market exchanges. It is this type of economy that has paved the way to so-called "mission-based" companies, which search for other types of performance apart from financial goals (the well-being of a region's inhabitants, the creative development, ecological and social innovation, etc.). We can see the value of this type of thinking for regional development.

In connection with this movement emerged the following:

---

29 Note that "competence", for the public bodies, falls under the scope of the right to do (*I may*), whereas in the private sector, it emerges from the concrete capacity to act (*I can*). Let us not elaborate too much on those who *may* but *cannot*, and vice versa.

– A self-managing governance system, within the framework of Elinor Ostrom's theory of the Commons. This is a good idea for small, localized systems where the low number of participants makes them operate by mutual adjustment[30].

– A governance model through mediation. This approach is very interesting for sites where there are many actors with heterogeneous statuses and whose contributions must be regulated and energized without an overly restrictive framework. It makes it easier for public and private actors to reach an agreement, while engaging, when necessary, partners useful in decision-making:

- experts whose opinion is important. It is still necessary to choose the best ones, those who have knowledge and can adapt it or call it into question (rather than prelates or preachers who impose their ideas instead of contributing to general reflection);

- or stakeholders less directly involved than the central actors (companies and public bodies) but having their say for specific reasons (as citizens, users or consumers). In the case of tourism, the main types of tourist clientele on the site are important stakeholders.

3.5.3. *The advantage of collaborative governance*

Most Localized Tourism Systems are heterogeneous blends of individuals and organizations. Governance may be intertwined on the same site depending on the structures present (private with economic goals, private with an associative status, public bodies).

But only collaborative governance can distribute initiatives, develop creativity and increase the collective intelligence of the systems considered[31].

As Trosa (2012) pointed out, public management would be less in crisis (of legitimacy or effectiveness) if it shared its action initiatives locally[32], providing

30 The actors still need to be trained to intelligently defend the "Commons" they are supposed to protect. This became clear during the summer of 2023, when the owners of the La Teste-de-Buch forest, which was seriously burned, were unable to agree on the maintenance of their hectares. And it is still not certain whether the community of owners of Grande Brière, in Loire-Atlantique, will easily agree to new techniques that could complement their traditional knowledge.

31 We can also refer to what Clerc (2020) has stated about it.

32 For over 50 years, public management discourse has advocated giving power back to the field, due to the proven inability of the center to define local policies on its own. *On ne change pas la société par décret*, by Michel Crozier, dates from 1979 ("we must defend the idea of germination, of the liberation of potentialities, the only compatible idea with successful changes). But the need for control and the fear of divergence (analyzed as a source of inequalities in the face of rules) hinder this public movement, which is all the same necessary.

inspiration and potential for achievement. It creates even more dysfunctions if it burdens existing governance through public control with multiple constraints. On the contrary, it is effective as long as it uses its regulatory powers to make its territory creative and facilitate collective actions.

Collaborative governance is all the more useful insofar as it favors the expression of diverse preferences and interests of the actors in the tourism system. If there is a mediation system, then neutrality is of the utmost importance, because it has to integrate actors having complementary challenges, or because it is coordinated by a body primarily oriented towards the general interest.

This type of governance is not a mere consultation. The stakeholders who accept the rules engage in a collective, formal and deliberative decision-making mode. The process tends towards consensus or compromise, requiring a minimum of trust, which will encourage the construction of shared understanding and responsibilities (Ansell and Gash 2007).

In Localized Tourism Systems, it is necessary to check that all actors are clearly represented. The type of governance chosen must be able to:

– respect the relative independence of actors, depending on the types of tourism systems;

– engage them in principles or rules of collective action giving them room for maneuver, while making it possible to have common projects and organize their implementation.

It takes time to build the system fairly, guarantee the quality of the members and accept its variable geometry. For example, it is necessary to welcome the newcomers deemed appropriate, integrate the different sectors of activity and ensure public/private balance. It is rarely very simple.

Balance on paper is not necessarily reflected by the balance during the sessions or their field-based extensions.

We must combine public action and private activity, while respecting their respective skills. In a tourism system, nothing compels regulations to be directed by the public body. Everyone may play their role (exercise their profession or their expertise) at the service of jointly decided projects, using the resources at their disposal. Rather, it is attainment, coordination and animation structures adapted to the local system which play a fundamental role.

Collaborative governance should not be considered as remediation against authoritarian systems at the hands of the public authority, shareholders or private managers who have seized the right to decide. We must focus on opportunities in terms of initiative, skills, inventiveness, responsibility, the understanding of problems and the consensual sharing of projects.

Governance can neither be authoritarian nor too lenient. The question lies in finding the "golden mean", keeping the system in appropriate tension.

3.5.4. *Dehierarchization and realism*

In *La décision stratégique en action*, Lauriol (1998) showed that there is no collective governance without the strategic capacity from all the participants. It is not possible to imagine it without actors trained in decision-making and belonging to active networks, as seen above.

Since Allison's reflections (1971), we now know that decision systems are a complex and unstable combination between:

– the tendency towards imperiousness of one or more local *leaders*;

– sometimes complicated rules of action, related to the institutional and administrative structures of companies or public bodies. In systems governed by legislation or organizational standards, we cannot avoid a certain number of procedures, legal obligations and action modes imposed by regulations;

– the ability of the partners involved in collective action to generate debates, exchanges and initiatives. They must be given the power to do so, and be mobilized in a satisfactory way. The more humility statutory leaders have to share their power for the benefit of the collective interest, the more the bureaucratic constraints are alleviated and governance opens up to the actors concerned.

While it is necessary to have coordination bodies, collective decisions must be considered collectively. Governance is only a means for stimulating and supervising (with the collective interest in mind) what will have been designed and deployed thanks to multiple interactions.

According to Lauriol, in systems that are not completely hierarchical, only a dynamic of participation and innovation may make it possible to associate local actors with relevant collective projects. In any case, this is what the socio-cognitive approaches to collective action have shown.

Even though it is useful to count with the support of expertise or arbitration, only the opportunities for sharing information, exchanges and discussions will make it possible to change points of view and develop innovative projects.

A common solution cannot emerge without the development of new representations, without the actors experimenting together with new ways of thinking and sparking creative behaviors.

While it is true that political action needs personalization (Monod 2012), these *leaders* have to be competent persons seeking to convince through reason, without manipulating through passion. These appointed or emerging leaders must encourage exchanges and the contribution of ideas, without deliberation running out of steam or going around in circles. At a given moment, their real power is to assume collective decisions, to push forward jointly designed projects and to put them into action.

Finally, collective governance avoids depending on external arbitration (e.g. on experts and influential political actors who are peripheral to the tourism system itself, while seeking to impose their interpretations and convictions on what should be done). These offbeat arbitrations sometimes contradict field experience, and disqualify the local actors, by undermining their capacity to take charge of themselves.

3.5.5. *Management of conflicts between actors*

The literature abounds on how to manage groups and make them optimal. The reader may refer, for example, to Ancelin-Schützenberger (1971) or Saint-Arnaud (2009). Both authors have insisted on the importance of reducing dissipative energies, those pushing actors to fight between themselves instead of helping one another, or those trying to jump ship from a tourist system because they no longer find any interest in it, or do not tolerate power takeovers or the inappropriate behavior of some of the key actors.

From the point of view of public actors and the communities they govern, here are some examples of what collective governance should be able to reduce:

– recovering local creativity, to gain value as an institution;

– increasing bureaucracy instead of reducing it;

– deciding at the place of field actors without having the necessary experience to do so, or without an appropriate model to produce efficient recommendations for action[33], etc.;

33 Unless they have been thoroughly trained in management, many elected officials believe that their political power can be satisfied with rules that they issue or disseminate to impose action, without the need for managerial methods or suitable organizational skills. However, collective action simultaneously needs political, legal and managerial capacities to act, as well as geopolitical, economic, social and societal skills to set action goals.

– promising aid, subsidies or support which are not delivered, and subsequently placing actors in the field in great difficulty (financial, strategic, etc.).

From the point of view of other actors, conflicts are often related to their personal characteristics, their prejudices and the way in which they represent reality based on their status or their workplace. They also result from real competition on economic markets.

When conflicts come from representations, we may modify, soften and reconcile them. If they are due to competition, trade-offs can be made, which depend on the nature and intensity of disputes. It is advisable to highlight convergences and common interests as much as possible. Competition is not only about customers or trade, but it also concerns the sharing of public subsidies, the support we will be able to obtain or not, the priority given (or not) to the activities we are responsible for.

Finally, conflicts may be related to rivalries or personal grudges. All of this must be soothed. In complex situations, there are never simple solutions, the main task is to get rid of relational roughness, blockages and energy losses.

When a local system is too undermined by dissipative energies and internal conflicts, one of the most effective solutions lies in the emergence of a sufficiently strong and caring leader:

– who can bring together energies, help build a project and overcome divisions or personal rivalries;

– who knows how to build governance open to all of the internal actors by facilitating their debates, networking and mobilization[34].

34 In the case of the Louvre-Lens, leaders *energized* all or part of the territory, but did so in partial areas or within too short deadlines. The mayor of Loos-en-Gohelle for the ecological boost of his town, the president of the Nord-Pas-de-Calais region for the relocation of the Louvre to Lens. In Amnéville, two mayors successively sought collective answers to the problems they came across. The first through an authoritarian but unifying paternalism, the second by seeking to overcome differences which appeared later, through the creation of a public company capable of building shared solutions. However, the network of initial actors has broken up, the management of many establishments is now off-site and it is becoming complicated to get them to join an enticing project. However, the place is still attractive as a conglomeration of tourist actors in close proximity.

3.5.6. *Governance evaluation within the Localized Tourism Systems studied in this work*

In the cases studied, we have come across a fairly French paradox. Governance is generally more active when directed by:

– private entrepreneurs related to local elected officials who welcome them, or facilitate their establishment;

– political figures transmitting creative impetus, shaking up existing structures and giving rise to successful sequences (Amnéville, Nantes, Louvre-Lens, Saint-Jean in certain aspects).

We would not necessarily need to have local excitement, active networks or innovative territories.

Individual power is limited in time and in the intensity of what it may produce. Without the joint creation of local systems by a majority of local actors, projects lose longevity, practical consistency and potential for innovation. It is essential to optimize both the overall framing and the open density of the territory's tourism initiatives[35].

We will discuss this in further detail in Chapter 4.

3.6. Acknowledgments

The authors would like to thank Madame Aude Ducroquet, lecturer in Management Sciences at the University of Angers, for her careful rereading of the texts and her contribution to the citations in this chapter.

---

35 This is what R. Blake and J. Mouton showed in *The Managerial Grid* (Gulf Publishing, 1964). All *management* must be guided by practical challenges (projects, active search for results) and the engagement of actors. Both of these aspects require a lot of energy. If it does not search for results nor to summon relevant actors, management is passive (this atony is observable in certain territories). If it concentrates only in mobilization or operational focus, it loses optimization. The articulation of the two aspects requires working on goals (operationalization, product or service innovation) and actors (network management, intermediation, managerial innovation).

4

Typologies and Recommendations

Chapters 2 and 3 interpreted the cases from a geographical and managerial perspective. This chapter attempts to establish a general classification of the situations observed for intervention purposes. The typology we propose is more than a nomenclature of what already exists. Instead, we seek to understand its variety and act with awareness.

Our work also aims to identify operational opportunities in each of the cases treated, considering their variety (history, location, territorial structuring, current projects, organizational methods and managerial practices). In the field of management sciences, we embrace the pragmatic approach (Lorino 2020), which seeks to implement "actionable knowledge" (Avenier and Schmitt 2007).

Every situation presents strengths and weaknesses. Only a diagnosis anchored in the territory and carried out by enough local actors will develop an appropriate strategy for the evolution or consolidation of the systems studied. In this book, we do not claim to offer solutions a priori, but to enrich the reflection of decision-makers in the territory, hoping that they will adopt open and creative forms of governance.

In the tourism economy, as elsewhere, established operations coexist with emerging models, whose actors contribute new projects to be embedded at the local level. For this, it is essential for them to learn to cooperate[1].

---

1 This type of analysis was theorized by A. Lipietz for the social and solidarity economy (e.g. *Pour le tiers secteur*, La Découverte, Paris, 2001). At one point in their history, the established social economy and the booming solidarity economy were wary of one another. Things evolved positively, but there are still implicit tensions, such as between tourism and culture, or between innovative centers and the actors of regular tourism.

How can we improve Localized Tourism Systems, be them well-established, emerging, dysfunctional or deteriorating? In an ample sense, we can find inspiration in the theory of strategic actors by Crozier and Friedberg (1977). The systems we are referring to are "concrete systems", struggling with an interplay of actors. These seek to consolidate their capacity for action, or at least be granted personal leeway. However, they only have a limited rationality, nourished by partial information, which is not always easy to evaluate. They proceed as best they can, despite uncertainty on what they should understand or be able to do.

Within these systems, by all available means (projects, incentive networks, facilitating governance), strategic work involves enriching limited rationality. We must transmit interpretation and action tools that will inspire desire – for the "good reasons[2]", in the words of Boudon – to increase solidarity or cooperation between local actors. We have to show that collaboration around certain topics proves to be more profitable than purely individual strategies. Contributions in the form of partnerships, knowledge and initiatives enrich the system with a better potential for action, regeneration or economic results.

To achieve this, we need a continuous and multidimensional work – even admitting hesitation or comings and goings – because there are always temporary withdrawals or moments of tension inherent in the flourishing of collective trust. We will see this further when we address the recommendations. However, as it is better to start with the local situation, let us begin with a classification of the forms that the tourist systems studied have adopted.

4.1. Two open typologies

In the following paragraphs, we will present two successive typologies of the cases studied, and then a series of recommendations. Firstly (section 4.1.1), we will look back on what systems are, and how they may acquire incomplete or unfinished forms. Then, we will emphasize the importance of proximity (section 4.1.2.1) and the need for openness (section 4.1.2.2).

The first typology (section 4.1.3) classifies the cases depending on the strength and consistency of the systems: with strong, diffuse, or emerging, tangled, deteriorated or merely apparent governance.

The second typology (section 4.1.4) focuses on the actors and their internal dynamics.

2 Refer to R. Boudon, *Raison, bonnes raisons*, PUF, Paris, 2003.

4.1.1. *Characterization of Localized Tourism Systems*

4.1.1.1. *Conditions required to "make a system"*

When can we really talk of a tourism system? When we can observe three interrelated phenomena:

– The interactions between the system's members are generally more numerous than the exchanges with the actors located outside, at least in relation to the tourist activity they are involved in. These interactions can be analyzed in terms of frequency and relational intensity. They include direct interpersonal exchanges or intermediation systems (meetings, institutional creations (semi-public and public companies, associations, etc.), events, third places[3]).

– Common goals or objectives can be identified. Although they are not necessarily formalized or completely clarified, they are proof of a global dynamism, as if the system had a collective soul without which it cannot exist, because it is not self-determined[4].

– Internal processes make it possible to operationalize goals. They are of two orders, both complementary and necessary:

1) a form of internal leadership that encourages the implementation of collective actions. This leadership can be driven by vigorous individuality (a charismatic and entrepreneurial political leader), effective plural governance or a mixture of both;

2) enough actors in the system who become engaged: tourism professionals, customers, employees, capital or project providers, suppliers, communities, public authorities. Engagement means that they commit a minimum of personal resources, because they hope to collect rewards at least equal to their initial contribution (money, developing or securing an activity, indirect resources, notoriety, power or political influence, etc.).

In some of the cases studied, this dynamic exists. In others, we are far from being able to observe it. We might then be surprised by the term "localized systems" when territorial realities end up functioning in a fragmented manner and whose finalization, organization and coordination means are fragile, and sometimes non-existent.

---

3 Possible forms of intermediation are briefly described in Chapter 3.

4 It is not because we inject contractualized goals accompanied by financial resources or political support (such as a destination contract) that this will lead to a project producing significant changes within the tourism system concerned. The destination contract is a useful adjuvant, provided that resources can be mobilized, and networks can accommodate the contract at the local level. It is not intended to fit into a vacuum of collective participation. This is also why it requires agreements signed between credible territorial partners. It needs to lead to collective operations transcending formal signatures.

Nevertheless, there are embryonic networks, and we always find forms of exchange between part of the actors. Even though the only apparent tie is proximity within the same geographical location, we can still identify what, under certain conditions, could serve as a basis for more efficient collective actions.

4.1.1.2. *Conflicts, controversy or uninspiring goals*

Within a territory, we must pay attention to what is missing or has deteriorated in the usual constituents of an open system (goals, networks of actors, boundaries and flexible governance). This is because these insufficiencies deteriorate existing systems or hinder their emergence when they are not yet established. For example, poorly resolved power conflicts hinder *leadership* and reduce engagement.

At the same time, we must be attentive to the goals proposed. These are only of interest as long as they help actors to gain strength (by enriching resources and knowledge), or if they give them attractive perspectives in relation to their profession or business system. For example, they can improve their practical skills or offer them innovative responses to market expectations or territorial requirements[5].

On the goals or methods of action, there may be divergent points of view (depending on professions, people, types of experiences or expectations). Differences in analysis often lead to interpretative controversies about what to do, why and how. Without finding a minimum harmony or acceptable compromises, it is burdensome to act together.

Back in Chapter 3, we emphasized that we must address both specific challenges and collective goals. Even at the level of the ethics of action (Morin 1994[6]), actors first have duties towards themselves, their loved ones and their direct groups of belonging (profession, socio-professional categories, etc.). We cannot ask them, without progressive[7] support, to put themselves directly at the service of their territory or the human community, even though they understand the interest or importance in doing so.

Furthermore, a Localized Tourism System cannot be permanently weakened by external upheavals that disrupt it at the slightest hazard (political change, economic

5 Manifesting as the well-being of local populations, job quality, and sustainable and eco-friendly tourism.

6 See the section on "complex ethics, between anthropocentrism and universal targets", pp. 352–356.

7 This is of the same order as Maslow's needs (*Vers une psychologie de l'être*, 1963, French edition by Fayard). If we do not satisfy our needs for survival or security, we cannot access the needs for relational sharing, recognition or development.

disorder, accident, technological threat, etc.). It must learn to face such disturbances and robustly defend itself[8].

4.1.2. *The role of proximity and territorial resources*

4.1.2.1. *Local potentials*

A tourism system is localized when it relies on proximity and territorial resources. It may be a specific place, a site, a town, an inter-community, a department, a regional area with specific historical properties or a geographical area with a recognized and territorialized heritage (Loire Valley, Dordogne Valley, the Vendée coastline, a mining basin, a vineyard, an attractive urban area, etc.).

This is what happens in most of the cases studied, even though we find boundary conflicts or the absence of demarcation (due to vague extensions or impacts outside the base territory). In France, a centralizing country, there are always national or regional power games that partly limit the autonomy of local systems (laws, general frameworks, imposed formalisms, subsidies subject to certain injunctions, etc.).

At the Louvre-Lens, for example, the museum partly depends on the Louvre-Paris and the founding initiative came from the regional president, not from the urban community where it is set. Beyond the impulse coming from the outside, it was the local interactions that triggered tourist development by mixing attractions and networks previously present on the site (mining basin, original landscapes, museum fabric, Flemish towns, belfries, war heritage, and football in Lens).

If they are well activated and regulated, agreements or local networks (both geographical, relational and productive complementarities[9]) amply contribute to territorial development, through their ability to act in consonance with local culture, based on properly understood and well-integrated issues[10].

8 Of course, this is partly the role of public systems. However, the system must also equip itself with resources (proactivity, projections, action skills, insurance). There is nothing more unrealistic than hearing tourism or cultural leaders declare that they are devastated by events whose effects they could have largely anticipated or reduced. Some complain about it as an unforeseeable catastrophe and demand that the state "take responsibility" (fairly recurrent speeches on media).

9 This includes service and logistics partners, or the actors making up the techno-structure; that is, all the devices facilitating production actions. This is related to the description in Mintzberg (1978) or Porter's value chains (1986).

10 This is why, in the theory of public action, we call for decentralizing decisions as much as possible by applying the principle of subsidiarity ("not taking power over what can be understood and decided in the closest proximity"). One of its founding articles was written by Jean Pierre Worms and Pierre Grémion, "*L'État et les collectivités locales*", *Esprit*, No. 1, 1970. It influenced the first decentralization laws (1980s).

4.1.2.2. *Territorial and inter-system extensions*

Within the same place and beyond it, every system needs to exchange energy and knowledge with other systems.

Even though it is predominantly local, a system must find ties with other systems in the territory or at a larger scale, depending on its needs in knowledge, industry and its various deployments (attractiveness, communication, development). It is under these conditions that it will renew itself and enrich its resources.

In this perspective, interclustering[11] is also a promising way. This is what could have happened in Cognac between the wine cluster (which had reached its maturity), and the attempt to create a tourism cluster within the same territory. Due to the lack of sufficient energy in the tourism sphere, it failed to be efficiently implemented. However, it could have been "an additional lever for actors, both private and public, because it connects potentially complementary economies while increasing available resources" (Cusin and Loubaresse 2015, p. 13).

In addition, it is necessary to note the strategies for expanding "private organization" tourism systems: Center Parcs has multiplied, Puy du Fou is trying to follow its example, the major attractions in Amnéville are today independent or managed remotely by investment companies. For its part, the *Voyage à Nantes* had its own spin-off as the "Dragon of Calais".

Despite the territorial distance, exchanges between sectors with the same type of activity are sources of mutual enrichment, as in the case of the mountain cluster, site associations and cities (the most beautiful villages, cities of character, art, history, etc.). In the France tourism lab network, feedback has been regular, and there have been learning expeditions (as in Montreal or Barcelona).

A priori, the coexistence of systems within the same territories should not be considered as a disadvantage. It all depends on transactions and complementarities. For example, the *Destination Anjou* association (hotel and catering) allows its members to cooperate smartly with other structures or systems ad hoc, benefiting from their contributions and actions, whenever it cannot find what it needs.

Instead of building public (heavy and formal) operations to develop tourism in Cognac (while being devoid of large budgets and facing a rich and powerful industry with international networks), the region could have acted based on complementarity and sharing mechanisms.

---

11 Contacts, exchanges and productive synergies between two clusters.

4.1.3. *Forms taken by the systems in the cases studied*

4.1.3.1. *Local systems with a significant degree of internal cooperation*

This is mainly observed in two types of cases:

– The tourism form is a business company with hierarchical actors engaging in projects decided by the management (Puy du Fou, Center Parcs). Most of the time, these projects turn out to be largely independent from public governance, even when they are tied to the territory. Some are "above ground" (expanding towards various places); whereas, others, despite vindicating a "local spirit", operate on a private basis that largely escapes territorial control[12] (note that the scenography at Puy du Fou is inspired by the history of the Vendée region and Parc Astérix by the myth of a Gallic defense against the Roman invader).

– There was strong enough leadership to unite and engage the majority of the local actors in the tourist area (as in the case of Nantes and, in its early days, Amnéville). This inspirational leadership (with projects that are both engaging and constraining) enabled professionals to truly benefit from it, even those who initially approached it with reluctance. Only then did they accept its relative disadvantages.

4.1.3.2. *Partial emergences*

Tourism systems have emerged sparsely, either in the embryonic form (the case of Lake Geneva, which is evolving positively) or through collective actions that failed to thoroughly accomplish the initial goals, or whose progress was hindered.

Current obstacles or past blockages result from three possible reasons:

– Public legislation (such as the NOTRe law[13]) and the forms of aggregation it imposed do not always match local needs. Despite an auspicious beginning, the Morbihan cluster did not survive. On the Vendée side, around Saint-Jean-de-Monts, rezoning was also a waste of time, as it created divisions that did not exist before that and introduced tensions in terms of action strategies.

– Conflicts between political actors or professionals in relation to their professions (for example, tourism and culture), or the overlapping of heterogeneous

12 At the time of the Covid-19 pandemic, Puy du Fou sought to negotiate specific agreements to loosen entry quotas to Paris. Also, it is also trying to outsource its attractive know-how as Disney or Center Parcs have done.

13 NOTRe, "New Territorial Organization of the Republic", 2015 (under the presidency of François Hollande).

territories[14], adversely affected the changes implemented in the *Saumurois*, the Dordogne or the *Autour du Louvre-Lens*.

– Tourism was not one of the territory's strongest points. Even though it has made some progress, it is still far from initial expectations. For example, it overlooked approaching more solid action structures, which (even facing more fundamental challenges) could have served to exploit tourist activities for the territory's benefit (as in Cognac).

4.1.3.3. *Conglomerates of tourist activities*

In various places, the territory is essentially home to an amalgam of activities with partial ties, without these being generalized, systematized, or designed with a view to optimizing reception or increasing the destination's attractiveness. There is often a "hoped-for system", but it basically exists in the minds of local decision-makers, who strive to find certain synergies, while having difficulty to strongly engage the actors.

Certain territories are essentially conglomerations of tourist supplies, brought together by a minimum of public regulation (municipal decrees on the area's functioning and facilities, the valorization of local assets by the tourist office, etc.).

It is always surprising to see the extent to which local authorities, tourist offices or institutional systems created on site adopt structuring discourses for their actions (often based on national models), rather than constructing these categories starting with the needs of local actors, as felt or expressed by the latter.

However, it is an essential condition for engaging them, provided that we simultaneously equip them with resources in terms of comprehension, possible solutions, territorial prospective and relationships with key players who can help them in terms of inspiration and opportunities. Without this, the initial energy of finally being consulted runs out of steam, due to a lack of creative fuel and reflections sparking innovation. We must constantly renew the desire to participate, as it is stimulating for the activity.

---

14 The geometry of territories depends on the destination (perceived by the tourists themselves) and local habits (for example, depending on tourist amenities and traditional forms of aggregation, in particular along a river, by the sea or in the mountains). However, administrative regions with sometimes discordant boundaries and contours are superimposed on them. To give an example among one of the best-known sites in France, the first thing visitors to Mont-Saint-Michel notice when they arrive in the half-empty building that was built near the shuttle boarding point is the rivalry between the Brittany and the Normandy regions.

Most of the cases classified as emergences (section 4.1.3.2) are found here, because they oscillate between a unitary display and a return to a certain activity disorder. Only Saint-Jean-de-Monts and the municipalities associated with it are currently doing well.

4.1.3.4. *Entanglements of systems and networks*

There are territories where system entanglements coexist (e.g. the mining basin where *ALL* is set). These systems have different sizes; some are strictly piloted, others less so, some are developing, others stagnating or disappearing. They interfere with one another in a complementary, tensional or unsatisfactory manner.

The question then is: what is the optimal type of governance? In political terms, should we develop general coordination governance for existing networks or systems? Should we stick to public prescriptions regarding territorial competences? Should we favor the most active system, encouraging it to take charge of a sort of overarching leadership, with sufficient degrees of freedom to leave the initiative and satisfactory complementarities to other networks? Is there no reason to consider existing systems as individual actors, articulating opportunities for cooperation and the need to remain competitive (or at least in emulation)?

Is it absolutely necessary to unify? Cannot we let adjustments take place through collective action, by motivating the tourist territory to become an innovative territory? There will certainly be competitive tensions, but this will be accompanied by a greater accumulation of resources (cognitive, practical, innovative) and a general production capacity[15]. We do not necessarily need all the actors, nor many of them right away (especially when the initial actors have disparate concerns, heterogeneous sizes and sometimes experience risky competition[16]).

Through this exploration, we come across the fundamental question regarding the necessary ties between an emerging Localized Tourism System and its innovative territory. We will try to answer this further in the text (section 4.3). Perhaps it is necessary to draw a distinction between at least two types of complementary actors:

---

15 It has been shown that in living systems (animals and plants), a territory rich in resources often increases internal competition without preventing the coexistence and development of competing species. Besides, it strengthens synergies and innovative behaviors.

16 For example, a tourist office wanting to make a buzz will sometimes have difficulty convincing the field actors who are struggling to make ends meet, or who are too busy with their daily activities. A persuasive effort requires facts, takes time, and cannot bring everyone together.

– proactive actors, voluntarily committed to a tourism deployment they help to reinvent or update;

– more passive or more routine actors, but whose productive contribution is needed at certain places or moments of the comprehensive project.

Public services tend to standardize and, in doing so, reduce variety. The private sector is utterly different in that it relies on the specification of competitive advantages[17].

4.1.3.5. *Systems which are degraded, in decline or running out of steam*

Finally, there are degraded systems (or others in the process of degrading). In practice, this is normal, since, by definition, these systems are structures that resist their disappearance, and whose maintenance, resistance or development energies depend on actors and contexts. As much as we can create them or make them grow, they can decompose, shrivel or die.

Some systems may have demonstrated appropriate coordination at one point in their temporal trajectory but have eventually weakened or become scattered. This may be related to the supply itself (the attractions suggested no longer correspond to the current needs of tourists) or to governance crises (projects are non-existent or unsuitable for the economic situation, exchanges between actors decrease, general management is deteriorated). This becomes evident when dominant and influential members have left, when conflicts have broken out, when the system has not succeeded in overcoming a financial crisis, etc.

Sometimes, the ties created were only superficial (e.g. gathering all the local supply under the same catalog). Over time, if there are no perceived advantages nor any prospects (useful differentiation or innovation), the actors no longer see the point in being a member. At the same time, professional networks have multiplied (types of sites, activities, brands, or professions), and this overabundance tends to disperse investment.

We must also ponder the question of available resources: workforce, products, financing sources (even in a territory where the key actors are elected politicians, the

17 One of the authors of the work was a member of PRES Nord-Pas-de-Calais, a university network bringing together six universities and around 15 schools. University presidents always wanted to standardize the training supply, remove what they called educational duplication, codify course contents and reduce the global training proposal. School directors laughed up their sleeve, knowing they would take advantage of this reductive standardization to gain in market share; as we cannot respond to an ample and heterogeneous demand with an a priori simplified supply.

risk of a public financial impasse is real; the only way is therefore to ally with private investors).

In the cases studied, we can observe three different examples of this relative degradation. A microsystem on a small territory, the Mélusine network seems to have gradually lost its own advantages, at least that is how its members perceived it. Most of them disengaged from it and acted in a more dispersed manner.

For Morbihan, the institutional structures changed responsibilities and dismantled the cluster project, even before it could reach maturity or find stability over time.

Unified	Tight conglomerate	Diffuse entangled
Actors in small numbers or in a restricted territory (leisure companies, theme parks)	Pools of activities obeying different directions (sometimes eccentric ones)	Overlapping or interacting pools of activities, with differentiated orientation and organization *leaders*
Actors pooled around a common policy, with intersecting marginals, and actors outside the system but in indirect relation with it (example: *Voyage à Nantes*)	When the territory is restricted, there are coordination obligations at different levels (e.g. Amnéville, Saint-Jean-de-Monts)	We may find (or not) global coordination (with federal leadership) (e.g. *Autour du Louvre-Lens*)

Table 4.1. *Systems: unified, tight conglomerate, diffuse entangled*

In Amnéville, the sudden disappearance of the founding mayor, the difficulties with succession and the pandemic crisis disrupted previously established networks (and all the more so as several key actors were replaced, and part of the management left the site). Having received assistance from the department, the current mayor is still struggling to put forward common goals and to highlight the project's collective interest.

To conclude this first typology, here is a brief partial synthesis of the forms observable when a local system is actually in place (based on the characteristics proposed in section 4.1.1).

4.1.4. *Distribution of cases depending on the types of actors*

In the cases studied, we can focus on the main actors, those who intervene to inspire action or influence general governance (section 4.1.4.1). We can also take into account all the tourist operators in the territory, paying attention to their evolution, collective dispositions and action energy (section 4.1.4.2). Both aspects offer *top-down* clarity on the available action potential.

4.1.4.1. *A characterization by key actors*

	Above all private	**Both public and private**
Rather focusing on tourism	Type 1 Center Parcs Destination Anjou Mélusine	Type 2 Amnéville Lake Geneva Saint-Jean-de-Monts TIL Cognac and Saumur cases are ambiguous[18]
Open to other activities (regional revitalization, social and societal issues, sport, culture, education, societal or ecological goals)	Type 3 It does not exist in the cases studied, but we find private actors who manage tourist activities by adopting other missions[19]	Type 4 Nantes Louvre-Lens Morbihan Dordogne[20]

Table 4.2. *Typology according to the public/private relationship and the degree of tourist focus of the territory*

Depending on the case, the dominant actors are private leaders, local public officials or a composite group with collective influence, with private/public regulations.

18 In these two cases, the aim is to strengthen the tourist impact by clinging to more powerful neighboring systems or by standing out from them. However, the actors are prisoners of complex geopolitical realities that complicate the situation.

19 The owner of Château de Brissac, for example, has invested in vineyards, event management and the hosting of film shoots; its park is open to residents; and its assets indirectly serve municipal interests. Along the banks of the Loire and its tributaries, there is no shortage of small castles combining several activities; for many of them, tourism is no more than a secondary activity.

20 In the four cases, although tourism development was presented as a distinct project, in practical terms, it interfered with other territorial challenges straightaway.

4.1.4.2. *A characterization of all of the actors*

If we look at things from the perspective of the basic actors (micro-tourist activities of a commercial or artisanal nature, small businesses in hotels and restaurants, centers hosting more or less collective activities directly or indirectly linked to tourism[21], etc.), there are several types of emergences that can (or cannot) profit from local leaders:

– Dispersed actors who, without institutional support, have difficulty building an attractive or truly profitable tourist system (e.g. the *Cité du livre* in Montmorillon).

– Actors who have made arrangements among themselves to build a functional system (e.g. sports activities networks in Corsica or in the Alps, organic wine producers sharing common cellars in Banyuls). It is not certain whether we can improve these systems through public intervention, or whether the actors will agree to this (especially if they are wary of the supposed risks of excessive influence, loss of control, bureaucratic burdens, etc.).

– Actors more or less assisted or stimulated by a large-scale public project with local *leaders* and support systems. However, there are also parallel risks of deficient economic reading, inappropriate means, discontinuous or conflicting ambition, and slowness in paying the volunteers who got involved (e.g. at the beginning of the Louvre-Lens in Morbihan, Cognac or Dordogne).

– Various actors partly integrated into a coherent tourism system with institutional promotion (in independent networks or attached to a central network, as in Angers, Saint-Jean-de-Monts or Nantes).

– Various actors who benefited from making themselves collectively visible in relation to an attractive destination (flagship destination contracts).

– Actors integrated into operations with strong interactions or in highly structured territories, but where the slightest collective problem can blow up the local organization (Amnéville).

– Salaried actors with unique private governance (Center Parcs, Puy du Fou, amusement parks, large zoos). It is sometimes gigantic, like the largest cruise ship, the *Harmony of the Seas*, which can carry 6,000 passengers; has a 2,100 member-crew; 8,000 people is the population of Saint-Jean-de-Monts. This type of attractive boat is nonetheless highly polluting.

In each situation, not only are the regulation modes different, but also the tourist effects on visitors (who shift from wandering or check-in self-organization to being captive in itineraries, circuits or pre-arranged stays).

---

21 Agricultural producers showing their estates, owners of small vernacular heritage sites open to the public, creators of craft objects available for sale, etc.

No matter the configuration studied – except perhaps in totally hierarchical systems where there are salary dependencies and organizational modes which resemble those of ordinary companies, such as Puy du Fou or Center Parcs – the actors have degrees of freedom and a certain independence from the system in which they take part.

As indicated in Chapter 3, they experience both cooperative and competitive relationships, a phenomenon known as "coopetition" in current managerial theory. The characteristics of coopetition have been fairly well documented. They also correspond to partnership development theories. Practices involve solidarity, complementarity, mutual support, and also competitive behavior at two levels:

1) power games in internal structures (for example, to influence collective choices or to coordinate public action;

2) that of economic competition when the activities target the same potential audience, who will choose between different forms of supply (catering or accommodation, tourist attractions responding to similar needs, preferences or expectations).

In terms of governance of a complex system, there lies a paradox: it is often useful to produce more unity (where there is a need for cooperation and shared goals) and more diversity (to broaden the supply, once we have agreed on the main goals and global process)[22].

For the trade-off between cooperation and competition to be realistic and promote mutual development (with advantages greater than the relative disadvantages), different actions can be tackled. We will describe them through general recommendations, whose flexibility is open to discussion. We can use what seems appropriate, in relation to each tourist context and each configuration of actors.

4.2. General recommendations

When you know that you do not know, you will know, but when you do not know that you do not know, you will never know (Fulani proverb).

22 It is in this way that living beings have become more complex by diversifying their adaptive opportunities, while acquiring behavioral consistency (even though there are conflicts regarding internal authorities in terms of decision-making or alternative courses of action).

4.2.1. *Questions of method, apologue of the Royal Navy and areas of intervention*

4.2.1.1. *Limitations of this work in terms of intervention*

Let us recall some principles from the analysis of organized systems (Crozier 1994; Friedberg 1994):

– The results of collective research do not lead to unique solutions. There are mixed reasons that explain the observed phenomena, such as cultural, institutional or local factors[23]. However, it is not the causes that matter, but rather the ways to act. The possible directions need to integrate the reactions of local actors and the goals they transmit.

– It is through collective behavior within an organized system that change can appear. To solve collective action problems, people first find support in the representations they have already internalized. However, they will be obliged to learn and innovate, from the moment that these cognitive patterns have become ineffective. Also, if we are interested in changing the system without denouncing potential culprits, the challenge is to find available resources, as much as possible within the system, or outside if necessary.

For his part, Barouch (1994) showed that case studies (or monographs) were forms of rhetoric that generate exchanges and controversies:

– Their rendition is one of the essential elements of a change strategy. If it does not tell the whole truth (or only tells it partially), it pushes the local actors to try to solve this discrepancy in interpretation, solely based on their habits of thinking.

– The case study favors certain identifications ("yes, something similar happened to me"), inviting us to correct our mistakes. On the side of local actors, it troubles misleading certainties and compels them to shift the interpretive lines, which sometimes makes it possible to unblock situations.

In this work, we have not been able to go as far as providing feedback to the local actors on the case studies carried out. However, the way we organize our recommendations tends to that goal. It makes room for confrontations, without claiming to state general truths or to propose actions of a universal nature.

4.2.1.2. *Apologue of the Royal Navy*

Concerned about certain decisions at the heart of action systems, Crozier (1979) judged that "society cannot be changed by decree". We must negotiate locally,

23 Routines, representations forged over time, power games.

through mediation and changes constructed at the foundation by the actors of each concrete system. In the context of public decisions, Worms (1994) believed that regulation is necessary on condition that it is enacted with caution and summons the relevant actors. This is possible through wise decentralization, by means of "decrees", which, like management tools, are means of supporting collective action rather than additional constraints.

Let us take a historical reference to understand how poor relationships between the public and private sectors can lead to perverse effects. When analyzing the stagnation of a collective action system (the Royal Navy at the end of the 17th century), Colin and Meschi (1998) emphasized the weight of structural inertia, strategic blindness and moral impotence, which prevented, slowed down or diverted a mobilization of resources that had already been decided upon (in the human, financial, technological and logistical fields).

At that period, when military needs were crucial, the highest political levels proclaimed "a great ambition" for the Royal Navy (Colbert and Louis XIV were highly involved). However, to achieve this, it was necessary to "rethink the traditional system for managing the workforce" (their salary was too low, the training insufficient, the professions not sufficiently attractive, the management inappropriate). It was also necessary to enrich the technical means, to find money and intelligently reorganize the arsenal ports (in terms of industrial structuring and supplies).

As already mentioned, the project did not successfully work out due to organizational inertia, as, on their own, "structural deficiencies are capable of generating low motivation and a poor morale, delays in decision-making, conflicts, and a lack of coordination" (Desreumaux 1992). Without a structure to accompany the strategy, we lose all effectiveness. Despite the historical gap, this analysis can serve as a comparative model for tourism systems, because the same causes produce the same effects.

In the case mentioned, there were communication problems and a poor distribution of resources. The management style of the time prevented any consultation on the field of action. In short, we were able to observe (what we see today in certain poorly managed tourism systems) an obsolete hierarchical framework, discontinuous and filtered information, conflicts of actors and a lack of education in a concerted implementation strategy.

Although this example may seem distant, human beings have hardly changed since the 17th century. It suffices to reread the moralists, in particular, La Bruyère

(1688[24]), on political cliques and other power games. Obviously, available resources, technologies and institutional schemes have profoundly evolved. Nevertheless, economic conflicts, tensions between territories and the multiplicity of wars remind us to be cautious in relation to certain a priori ideas on modernity[25].

4.2.1.3. *Possible areas of intervention*

After analyzing the cases, we identified points to be assessed in four interactive areas: tourist resource management (section 4.2.3.1), project implementation (section 4.2.3.2), the development of interactions in networks (section 4.2.3.3), and the engagement of all actors (section 4.2.3.4).

It will be necessary to add the consideration of general issues that mainstream the previous themes, and study them at a later stage: the importance of looking at the territory as a destination for tourists, of acknowledging their interests and representations (section 4.2.4.1); the necessary balancing between public/private, social/technical and managerial modalities (section 4.2.4.2); the incentives to train and innovate (section 4.2.4.3); the evolution towards sustainable tourism (section 4.2.4.4).

4.2.2. *A diagnosis phase by the actors themselves*

4.2.2.1. *The internal diagnosis*

Preliminarily, we must take some time to consider the diagnosis carried out by the local actors themselves. They need to be lucid about the history and genesis of their tourism system, regardless of its condition[26]. They must identify its strengths and weaknesses, mainly in relation to existing resources, the presence or absence of dynamic projects, the characteristics of tourist networks and their governance. It is from there that they will be able to consider other operations, other tourist practices and, if necessary, possible extensions to their potential for action (in terms of places, spaces, types of routes or ecosystems related to specific audiences and particular challenges inherent in the territory).

---

24 "The Characters, or the Manners of the Age", especially Book VIII, "Of the Court".

25 In contrast, Vauban the builder was able to implement collective efficiency on every site by installing temporary local systems with strong synergy. He is recognized for his attention to staff, his sense of justice and his concern to impose charters of good conduct on contractors in relation to their subcontractors. At the end of his life, he experienced the disgrace of the king for having suggested that he pay taxes like everyone else.

26 History has left important traces and behavioral habits that should not be neglected (collective languages, routines, know-how and also prejudices as a result of past experiences). Heritage must be actively evaluated (what shall we leave aside after the effort required to achieve it? What shall we keep or develop from what has been acquired?).

This need to localize the diagnosis (including its methods) is contradictory with a generalizing form of public governance.

For example, Atout France offers action models advocating an a priori classification of the actions to be carried out, without necessarily taking into account the territorial logic and its priorities. Not everything has to be done everywhere, and there are more specific challenges to a concrete territory than what the authorities at the national level are aware of.

We have known for a long time (Hatchuel and Weil 1992) that management tools do not exclusively have a technical substrate. They also carry ideologies (on what they will be used for) and representations of the systems to be managed (the actors involved, their interests and the forms of organizations chosen). The tools carry with them biased and politicized interpretations with a restrictive impact. At best, they must be transposable, transformable and subject to the evaluation of those they are addressed to.

Even in science, we are victims of nomenclatures (categorizations or systems of classification of reality) that impose their reading of the facts, by neglecting or repressing what poses a problem, and by emphasizing the points listed or supposed to be important (Martineau 2017).

Nomenclatures (e.g. "categories of actions to be carried out") have perverse effects if we consider their normative cartography as if it were the real territory. However, the map is never the territory; it is only a partial way of describing it for subjective reasons. Ultimately, we could arrange for each tourist to shape their own map of the destination they are visiting, based on the objects, situations and events they have chosen to explore, meet or consume[27].

In what follows, our proposals are developed as open diagnostic and action tools, which can be taken into account or left aside, depending on our priorities.

This diagnostic exploration must be done collectively, without deeming certain actions useless on the pretext that they are embarrassing, unusual, difficult to deal with or likely to lead to political debates that would nonetheless be necessary.

---

27 For a long time (1975–2000), the city of Chiang Mai in the north of Thailand had a very picturesque map, illustrated and regularly updated, to be used by foreign hikers. Embellished with small humorous drawings, it was richly documented for cultured globetrotters, lovers of temples, of pleasant sites, of curious environments and, of course, hotels, restaurants, tea houses, markets, boutiques and massage parlors. An ordinary resident of the city might not have found it particularly interesting.

4.2.2.2. *The contextual diagnosis*

Strong points	Weak points
A good number of flagship destinations with prestigious character. Great cultural and climatic diversity. A variety of sites, landscapes and terroirs. Rich heritage. Numerous facilities (museums, parks, monuments, etc.). Good dynamic of artistic and cultural events (festivals, exhibitions, events, shows). Recognized know-how and outstanding reception skills.	Lack of budgetary allocations for tourism prospection[28], sector engineering and international promotion (despite recent efforts). Not always appropriate[29] training resources. A shortage of staff (employees who are not sufficiently recognized, remunerated or involved in decisions). Relative aging of facilities[30]. A sector that should innovate more, including its products and services supply.

Table 4.3. *Strengths and weaknesses of French tourism*

There are two levels of contextual diagnosis:

– The one, more circumscribed, which describes the opportunities or threats of nearby environments (beyond the tourist system itself): for example, employment or population problems, accessibility influencing the territory; political developments affecting the management of public resources (financing, orders of priorities); infrastructure changes with probable consequences on tourism (construction of a bridge, new motorway exit, development of a lake, closure of a site deemed dangerous, etc.).

– The broader one, which involves major challenges having an impact on tourism in France[31], and therefore, on a number of local systems. Table 4.3 presents a summary grid[32] of the main strengths and weaknesses attributed to French tourism.

---

28 This prospection helps to develop action scenarios for reading the profession's general movements, focusing on new practices and their effects. We are not forced to obey *mainstream* injunctions; we can be singled out while scrounging useful opportunities along the way.

29 Technological adaptation (digital or material type) requires skills that must have been acquired. This is the only way to make a creative use of it, not just passive or minimalist.

30 Material improvement sometimes has tremendous costs. It is better to check the funding we receive by accessing the categories of public aid or subsidies beneficiaries, or by embracing private investment programs.

31 To identify these issues, we can rely on national studies carried out by professional associations, by consortia of tourist sectors and activities, or by Atout France, *ADN tourisme*, UNAT, etc.

4.2.2.3. *A need for progressiveness to find meaning and build performance*

In a Localized Tourism System, the necessary changes must be made little by little, with certain caution, relying on the growing confidence of actors. It always takes longer than we would like. Some projects can evolve very quickly, but others need to adapt to the pace of people, without imposing unnecessary disruptions that could slow down the process[33].

Trust never comes from a structured general discourse, but from reliability acquired through collective action. If we make most of the actors execute what we have thought without eliciting their opinion, they will experience this cognitive injustice negatively. Too many leaders wrongly imagine that the conception of things is reserved for an elite, and that the rest of the players are merely expected to put into practice what has been decided at their place. To counter this, the dependent actors can either withdraw or curb decisions.

While it is true that we must give meaning to collective action, this cannot be done, no matter how. This is not a way to justify the biased choices of a few. Sense must be constructed by several persons, and constantly submitted to collective criticism or control.

According to Hatchuel (1997), any philosophy of action oscillates between the concern for reality and the formulation of an ideal project for the future. Men need an engaging representation of the activities they carry out. They use rational myths, in the manner of living metaphors that help them rethink reality and give it a future; at the same time, these provide them with shared meanings.

4.2.2.4. *Particular attention to crises*

Three different types of crises can affect tourism systems. Locally, it is appropriate to monitor them, to avoid being in reactive mode or responding too late[34].

32 In order to compile it, we summarized around 10 specialized websites. The strong points demonstrate the abundance and quality of French tourist resources. The weak points emphasize the inadequacies in terms of maintenance, heritage development and training of the actors.

33 For example, unnecessary structural changes, unjustified bifurcation of goals.

34 Part of the difficulties observed in many of the cases studied is the fruit of a delayed response to the territory's problems. On the contrary, the ingenuity of proactive *leaders* can gradually succeed in transforming the situation, or even in making it a model of innovation (Nantes, Amnéville). Each territory has its contingencies, which can be transformed into strengths (competitive advantages): the black color of the slag heaps in *Autour du Louvre-Lens* is becoming a magical and powerful color, as it was in the paintings by Soulages.

The first corresponds to exogenous threats, whose effects have to be limited and from which we must protect ourselves collectively. The recent pandemic, terrorist attacks, isolation or distancing, forest fires or the sudden increase in energy costs are only a few current examples.

However, there are also medium-term risks against which local systems have to develop innovative solutions (e.g. the demands of the "ecological transition", whose effects on tourism are sometimes direct: the rise of the seas and the scarcity of snow in the mountains requires solutions that will, in the long term, be far more transformative than the option of snow cannons or campsite relocation).

The second are above all endogenous. They are related to:

– internal deficiencies with dangerous consequences, especially when there are too many or not enough tourists, because tourist reception and its profitability come from an adequacy between supply and demand, even with complex variations over time and space[35];

– actors' problems that disrupt collective action or prevent the design of projects. This is either because they conflict or because they dissociate themselves from the common project due to divergent interests.

The third comes from the overlapping of tourist territories, whose tensions are caused by boundaries, public jurisdictions, image games and overall visibility. Sometimes, it is the proximity of a very touristy area (a "major destination") that makes it difficult to develop a territory and turn it into an autonomous tourist system, with a project of its own.

4.2.3. *Resources, projects, interactions and engagement*

First, let us return to the four operational levers of Localized Tourism Systems, as analyzed in Chapter 3. For each of these, we can act on several dimensions at the same time. These are briefly described in Table 4.4.

---

35 On the Vendée coast, in 2022, a buyer of campsites interested in investing and immediately making profits (by increasing their range, and therefore their cost) relied on an incompetent manager who did not know how to manage these places and ended up having to cope with unhappy customers (with legal action, temporary closure and site on the Internet).

Resources	Projects	Networks	Governance
Accumulation	Realism	Density	Inspiration and training
Renovation	Proactivity	Interactivity	Cooperation and collective engagement
Renewal	Innovative solutions	Cooperation	Territorialization
Transformation	Engaging and attractive values		Conflict management (to reduce blocking effects or nuisances)

Table 4.4. *Levers for change based on the constituent elements of the systems studied*

4.2.3.1. *Management of tourism resources*

Where tourism systems work best (for example, as in Nantes), we pay a sort of creative attention to the resources offered. It is often preferable to expand their allure, taking into account not only tourists, but also business travelers (individual or in groups) and the territory's population:

– It is first important to have enough, and if necessary, through a development policy. We can even build resources without marginal cost by showing monuments or landscapes temporarily left aside by concrete or commercial valorization ventures. At the same time, this accumulation must find an order that guarantees the complementary supply or upgrades essential contributions (for example, in accommodation potential). We may also want to welcome customers with heterogeneous needs in terms of tastes, ages, income levels or reasons for attending the place.

– We can restore, develop or renovate available heritage assets by giving the m an innovative presentation or use.

– We can bring new, original attractions, enabling us to make the territory stand out. Just to give an example, a town cannot be differentiated only by a castle or a church, because these are fairly common heritage sites, unless they are singled out or complemented.

– It is also necessary to transform certain resources according to new expectations. Take the case of Saint-Jean-de-Monts, a resort for families, where a seaside promenade is being redeveloped by moving cars away (deemed noisy and polluting, or dangerous for children).

– It is also in our interest to build circuits, itineraries and ways to discover the space through movement (on foot, by bike, by car, by public transport). In doing so, we establish active "wandering" and motivate cultural "traveling". We grant a certain freedom to visit or not, to find and to choose[36].

– Tourists need stories around these resources, because it is through *storytelling* that these grasp the voyager's imagination and are kept in memory.

4.2.3.2. *Project implementation*

Any project must always be considered as a form of voluntary movement to escape *path dependency*[37], the habits we have adopted and lock our actions into repetitive patterns. This gradually steers us away from the demands of reality and the evolutions they impose. We confine ourselves to it all the more while we are not overly under pressure (at least for the moment) by competitive threats.

From this perspective, projects can be tested on three characteristics:

– Are they proactive? We will not mobilize solely based on the need to defend ourselves or maintain things as they are.

– Are they realistic? We will not engage professionals on goals with a low probability of achievement, because they will refuse to invest money, time, skills or material resources.

– Will they make it possible to develop creativity by seeking innovative solutions capable of transforming the existing reality?

Beyond their practical rigor and the concrete opportunities they offer, projects always benefit from being expressed through illuminating metaphors, as previously illustrated through the idea of rational myths[38].

Finally, since a tourism system is a form of organization, its projects should also embody commitment values[39] for local actors. Expressed through institutional or

36 On the model of recurring articles in the French newspaper, *Le Monde*: "One city, desires, five possibilities". We can confront the tendencies of the texts' authors who play with the potential resources of places as if they were rambling influencers. The April 2023 edition features Montluçon and the art of mixing desires: cuisine (potato *pâté*), heritage (the medieval town), nature (the gardens), the industrial past (the large towers), and museums (the museum of popular music). So, we taste, we wander, we walk, we stroll (verbs used in the text).

37 Path dependency refers to a phenomenon of inertia when it comes to decision-making (whether by public or private decision-makers). By persisting in choices already adopted, we deprive ourselves of other solutions that could be better. Theorized by David (1985) (path dependency), the concept was first applied to businesses and then extended to public policies.

38 Hatchuel (1997); see also section 4.2.2.3 and footnote 44.

commercial communication, these values have an attractive power over tourists and other travelers interested in the territory.

4.2.3.3. *The development of network interactions*

In existing or emerging networks (or in those which could arise), it does not matter whether there are business, project or territorial strategies. Each has their own language and needs. For example, a network may involve activating local proximity, increasing professional technical skills, mixing or bringing creative intentions to fruition. Also, it is not necessarily the same people who will distribute energy everywhere.

In tourism, the quality of networks can also be analyzed as a local welcoming force for the benefit of those who visit the territory and interact with it in the form of sightseeing, accommodation, catering, access to sites, cultural participation or commercial transactions.

The strengthening and densification of networks depends on their allure, which encourages people to be part of them or to actively intervene in them. It is their multiple weavings that produce innovative environments. This owes to:

– their dynamism, the opportunities they present, the quality of the projects carried out there;

– the forms of governance that channel them, organize them and make it possible to increase cooperation by reducing conflict;

– intermediation schemes that enable these networks to function collectively beyond interpersonal exchanges.

We must take into account the time availability of those who join such networks. Some can engage at any time; others have more seasonal participation and still others have very heavy constraints at certain times of the year[40]. Due to this, we must find appropriate moments for exchange and cooperation.

39 It is not possible to describe the vast theoretical field on values here. Let us follow the approach by Rokeach (1973), who differentiated terminal values or "desirable end-states of existence" (e.g. equality, beauty, pleasure) from instrumental values, which refer to "modes of conduct" (e.g. courtesy, helpfulness, honesty, good humor).

40 Those who take part in recurring events (such as festivals) are well aware of this varying engagement, depending on the needs or professions.

4.2.3.4. *An engaging governance*

Whatever form of governance is chosen, it must be inspiring. It must help build solutions or perceive opportunities. In this sense, innovation is a means for improving the tourism proposition, while opening up new perspectives.

A positive trait is the presence of charismatic leaders suggesting ideas or directions of action. However, if most of the actors in the field are left out of prospective reflection or deprived of operational suggestions, they are cut off from their intuition or desires for the future. This leads to a general slowdown due to a lack of perspective or knowledge contributions helping us to renew representations.

Governance must encourage the engagement of as many people as possible in collective actions and projects. Often, dynamization is hampered by two opposing sources of dysfunction: the multiplication of standards or controls, as well as incessant changes, destabilize collectives (Thiéry and Sénard 2023) by causing untimely ruptures in *leadership*, operational structures or political institutions.

Cooperation is not natural, at least if it wants to achieve a proper level of efficiency. We must take care of the actors or collectives[41], both employees and public agents integrated into organizations, together with managers and autonomous workers (community executives, craftspeople, retailers). We must convince them that they will benefit from a certain degree of interdependence, and that this will not prevent them from maintaining their autonomy (Everaere 2007). Of course, they keep the activities they are responsible for under control, but they must also take into account the fact that they contribute to the results of the collectives they participate in, and that these results have repercussions on their activities. We can propose sharing a minimum normative framework to achieve greater common efficiency ("destination" policy, coordination of certain operational processes, complementarities of services, collective response to economic risks).

In that sense, in tourism systems, it is not only the private companies who have neglected their staff to the point of making them passive, contributing little, or failed to reward them with sufficient retribution (salaries, psychological recognition, leeway), so that the system can flourish sustainably.

Regardless of the tourism system, the construction of collective action frameworks should be flexible, concerted and susceptible to adjustments whenever

41 In the sense that we take care of the actors present in the system, because they are essential resources. We do this, for example, by providing them with information by offering them training adapted to their needs, by engaging them in exchange mechanisms that strengthen their skills or stimulate their perspectives, by helping them find solutions to certain problems, etc.

the situation requires it, both to react to threats and to seize opportunities. This involves regular discussions, with mediation actors to facilitate the process. Public bodies rarely leave to their agent collectives enough room for taking initiatives. There is a lot of agency control, forcing compliance with orders. Instead of helping their private partners, certain public agents impose on them the rigid constraints they themselves were burdened with.

In order to coordinate its territory of action, governance must implement mechanisms for cooperation, creativity or operationalization. This is also how industrial clusters work, with the goal of organizing productive complementarities, articulating general logistics or having concerted innovations. However, when it comes to optimally adjusting a variable geometry[42] supply system, tourism is built less on closely intertwined assemblages – whose articulation requires technical meticulousness – than on collective flexibilities.

We must also give ourselves the means for opening the territory to external ideas or resources, while avoiding centrifugal leaks from uncommitted actors, or structures following orders outside the tourist system. The cases studied have shown us several examples: public authorities outside the territory that impose their views, or private investors who control their tourist structures from afar (zoos, casinos, franchised attractions, campsites, hotel-restaurants, etc.) disposing with regular consultation with the local system. Governance must then find forms of rewards or constraints that demonstrate the interest of local integration and contributing to proximity consultation.

Let us return one last time to the need to manage conflicts, even those scattered or sotto voce. If we do not reduce them, there is a considerable loss of energy, especially if they indirectly cause dysfunctions. Box 4.1 provides a list of commonly observed governance errors or tensions.

42 Therefore, operators have to be polyvalent, and sometimes the same operator has to show extreme flexibility in the organization of activities. For example, in a seasonal system, certain employees can have different roles in function, depending on the presence or absence of tourists and their number, and seasonal workers must be integrated in any case (to a different extent, depending on the profession). A hot air balloon owner is subject to very unstable weather, which can lead them to decide on a flight at the last minute (and to warn the customers); as there are several hot air balloons, they must ensure that qualified staff are capable of boarding passengers without any risk, and that valets can collect customers at the imprecise locations of their landings.

Errors in analysis and goals. It can also be a question of persistent errors due to misdirection or beliefs that are blocking reflective openness[43]. A veneer of action models that are not adapted to the context nor discussed with the local actors.

A lack of engaging entrepreneurs. Tensions between decision-makers (diverging interests, conflicts of power). Too many control phenomena and hierarchical distances. Difficulties in building collective projects (finalization problems).

Insufficient common representations and shared trust. Not enough support schemes, regulation mechanisms, events fostering conviviality (coordination problems).

Technical inadequacies, skills deficits (including among those piloting collective action). Too much bureaucracy. Devices that underestimate the complexity of implementation or the time needed (and which end up getting boring). A focus on day-to-day activities (know-how and priority issues).

Box 4.1. *Management of risks, errors and types of recurring conflicts*

4.2.4. *Territorialities, balancing, innovation and sustainable tourism*

In this last section, we will address four interdisciplinary themes, which should be pondered in relation to the above. We will present them in turn. Table 4.5 summarizes the key points.

The territory as a tourist destination	Internal balancing	Incentives to train and innovate	Evolution towards sustainable tourism
Starting with tourist practices	Between public and private	Learning territories	Harmony with the inhabitants
Articulating territories and destinations	Inside the managerial levers	Innovation from tourism itself	Ecological emergencies
Tourism as a territorial planner	–	A collective intelligence effort	Tourists' education

Table 4.5. *Interdisciplinary means of action*

43 The works of C. Morel on *Les Décisions absurdes* (I, II, III, Gallimard, 2002, 2012, 2018) deserve to be read and reread by all decision-makers.

4.2.4.1. *The territory as a destination*

The tourist destination cannot be determined by a decree. Contrary to what poorly controlled appellations claim (destination X, destination Y, which are multiplying), the tourist destination does not necessarily correspond to an institutional territory.

Moreover, the territories themselves, in their definitional entanglements, are subject to sometimes excessive extraterritorial influences. In concrete terms, there are at least three types of tangled spheres of influence:

– The national sphere, related to international evolution in tourism. Why not inspire, stimulate or support, but not in view of ruling or deciding at the place of actors who hold local knowledge?[44]

– The territorial sphere, which is sometimes larger than the tourist systems observed (following, for example, the encroachment of regions, which have become larger and whose presidencies are sometimes far from some of the spaces they govern).

– The professional microspheres, which have their own requirements alongside those of the territory.

So, how do you determine what a destination is?

4.2.4.1.1. Delimitation of destination perimeters based on tourist practices

Fundamentally, tourists ignore territorial boundaries, with the exception of the national level, which is essential, notably in the tour operators' supply. Circuits crossing several states are a rarity. In other cases, the destination is first imagined by social individuals, an expression which takes into account collective representations, but also acknowledges "discrepancies" between persons, according to the approach proposed by Lahire (1998). It is a lived space, built on a daily basis according to desires as well as the constraints encountered, and the seizing of opportunities and suggestions made by the actors (as developed in Chapter 2). This fact is explained by the fact that the infra-state boundaries are less significant, their crossing does not require a visa or any particular authorization and also because the knowledge of the places by tourists is vague due to the distance that separates their residence from the

44 With far less serious consequences, certain zealous officials, who know everything about everything, act like those who, after occupying Tibet, demanded planting wheat instead of barley, whose yields were considered lower; except that wheat depleted the soil and barley was more resistant to cold (this was local knowledge from experience). The grain deficit caused famines and numerous deaths.

places visited. It follows that the scope of the destination must be based on what tourists first imagine, and then on what tourists do (see Chapter 2).

However, the latter include places in their projects and in their accomplishment according to different configurations (Ballina 2021):

– the hub, where the tourists stay and from where they organize day trips;

– the stopover, where the tourists pass by, without staying, as part of a circuit;

– a single destination, where they spend all their available time, including overnight stays;

– an excursion place from another tourist hub.

Therefore, two approaches should be carried out jointly. On the one hand, it is appropriate to start with elementary places, as perceived by the tourists, and not necessarily municipalities; in other words, basic territorial units. On the other hand, from these, it is necessary to identify the circulation between the circuits that unite the places and the stay hubs from where excursion-type links are established.

4.2.4.1.2. Articulating territories and destinations

As there is no necessary coincidence between territories and destinations, the actors necessarily end up in coopetition scenarios. It is the tourists who, through their mobility, suggest the relevant blending of places. Since the legitimate actors who develop policies and the workforce implementing them act in function of the territories (it is their role, after all), cooperation must make it possible to respond effectively to tourists' expectations. This approach will not prevent competition from remaining active. It is even stimulating to achieve the goals set.

This articulation also depends on the diversity of tourists, in the double sense of their collective identity and disparities. Different societies share representations of the visited other. The greater the horizon of alterity, the harder are these visions to understand by the inhabitants and professionals of the places frequented.

Furthermore, the increase in distance has the effect of extending the space visited as well as the gap between visited places. In function of their society, tourists identify places for their project, whose stopovers are all the more distant from one another, as the time constraint offers them a broader choice. In addition, every society is socially stratified, which multiplies representations depending on various categories.

Finally, distance and alterity imply that the more the tourists' skills and dispositions will be elicited, the more significant they are. See the use of space technologies (Stock 2008[45]).

These are functions of alterity. The higher this is, the more individuals appeal to such functions.

Institutional actors benefit from being in contact with foreign tour operators. The "Welcome to France" event, organized each year by Atout France and by regional partners, invites them to promote mutual knowledge. In certain situations, non-professional mediation can be exercised. Take the example of Chinese youth who, in Europe to complete their studies and who master at least one language, make it their duty, out of filial piety, to help and guide their parents (L'Hostis 2020).

The match between territories and destinations will therefore not be a single model but will be obliged to adapt to the markets. The alliances will be of variable geometry, formed to comply with the targets. At the same time, we have seen that proximity is necessary between professional actors for them to get involved and that a degree of mutual knowledge is essential. The regional scale seems relevant, provided that this appellation refers to "tourist regions" (Piriou 2019), instead of administrative regions (especially since the 2015 reform, which considerably expanded some of them).

4.2.4.2. *Balancing needs*

Let us now return to the reflection proposed to the actors of tourism systems from the point of view of supply, with expectations in terms of balancing (section 4.2.4.2), innovation (section 4.2.4.3) and sustainable tourism (section 4.2.4.4).

4.2.4.2.1. Public/private interactions

Since its very beginning, the state's vocation has been to "bring order to the world, by producing cognitive structures to reflect upon it" (Muller 2015). However, it too often achieves this via domination effects. The need to regulate social fragmentation does not exhaust the energy of the actors and the meanings they build together at a local level. To maintain coherence, there has to be a to and fro between sectoral public policies and the territorial creativity of social actors. This is because it is in the reflexivity of local initiatives that the different ideas about the world are elaborated, and, in doing so, social evolutions can be transformed.

45 "Mediation elements [...] which aim[s] to solve problems of space [...], distance, access [...], alterity" (Stock 2008, p. 157).

As Bartoli (2012) pointed out, the public sphere can no longer govern through abusive general standards, ignoring local challenges. It is itself "exploded, heterogeneous and reticular". There is no longer any state omniscience or noble political vision that can impose its rules on those responsible for putting them in action. We need to "reflect upon collective goals together, instead of implementing management tools without hindsight". "Tools are only useful as long as they are accompanied by a certain leeway". To conclude, we must avoid "unthoughtful mimicry" or the mere transposition of poorly digested tools. It is in local effervescence that we find the energy to act, not in a priori programs.

In general, public impetus is stimulating when it supports collective action (by helping, providing resources, channeling and facilitating practices), once having established a diagnosis and a concerted strategy.

Instead, it is harmful when it imposes directives in an authoritarian or bureaucratic manner, or when it makes promises it cannot keep. In that case, it disrupts the actors rather than engaging them. Concretely, public bodies must rely on local energy. According to Cautrès (2023): "There is a real political demand in the country, looking for a compromise between a listening policy and an efficiency policy". It is by activating both levers simultaneously that we can renew public action.

However, there are always mismatches between the public challenges on the common good and the partial interests of local actors, including political ones.

To take inventory and reduce them, Box 4.2 can serve as a checklist (Dumont 2012).

– Are relations between elected officials harmonious or conflictual (locally, with supervisory environments, or nationally, with "higher public bodies")?

– Is public governance centralizing, cooperative or lax (with scattered contributions)?

– Are the relationships between appointed officials and economic and social actors loose or synergistic?

– Have local actors learned to work as a team? Are we seeing common goals emerging? Are we able to mobilize "a bundle of converging energies"?

– Do the projects implemented respond to most of the territory's essential challenges? Do they include employment targets? Did they rely on territorial prospective?

– Are there any transforming *leaders* (capable of inspiring, promoting, energizing, generating support), frontrunners or networkers?

Box 4.2. *Nature of the interactions between public and private actors*

4.2.4.2.2. Managerial regulations

For collective action to be effective, the technical and social aspects of management have to be articulated. The socio-technical model was undoubtedly one of the most relevant managerial innovations of the 1970s–1990s, with considerable advances in industrial democracy and participatory management[46]. This was integrated into the strategic approach of organizations. These reasonable modes of operation were replaced by the return of managerial authoritarianism, following the monetarist revolution in the economy and the control of financiers over organizational processes. That was reflected by a return to Taylorism disguised under the terms "total quality", "choice rationalization", *lean management* (with an arbitrary downsizing of the workforce) or "externally decided restructuring, without knowledge of the field".

In territories subject to several results criteria, where financial resources are needed but without being an end in themselves, attention to socio-technical regulations is essential. Actors in the field must understand technologies and organizational choices, in order to organize themselves accordingly (and without having models they do not understand imposed on them). If we are convinced of the desirability of a technology, it is because there are means to explain it, transmit it and facilitate its appropriation.

Regarding the distribution of managerial functions, authors such as Tabatoni and Jarniou (1975) or Adizes (2004) have shown the need to articulate several types of skills: finalization, organization, production and control. On the one hand, we must know how to harmonize them with each other; otherwise, the management system would be unbalanced. On the other hand, we must learn to distribute them, because a single person does not have the personal abilities or the time to take charge of all of them.

Let us give some examples of imbalances to be corrected. An imaginative and proactive entrepreneur can let go of organizational constraints or coordination requirements. These management flaws are not always perceived at the beginning, when the impetus is still strong and certain obligations can be postponed or neglected. However, there comes a time when these have to be addressed. This was the problem in Amnéville, where Doctor Kiffer's successor had to struggle with legal irregularities, maintenance defects, equipment functionality, as well as problems of strategic competence among those who had been recruited at the service of the former administration.

---

46 See, for example, Ibert, J. (2017). Frederick E. Emery and Eric L. Trist. *Des systèmes sociotechniques à l'écologie sociale des organisations*. In *Les Grands Auteurs en management*, Charreire Petit, S. and Huault, I. (eds). Éditions EMS, 97–112.

Complementary categories in management	Adizes model	Tabatoni–Jarniou model
–	Entrepreneur	Finalization
–	Organizer	Organization
–	Producer	–
–	Facilitator	Coordinate
–	–	Control

Table 4.6. *Types of managerial skills*

On the contrary, an organized system that functions but has no inspiration or creativity loses its entrepreneurial potential and jeopardizes its finalization subsystem. It risks falling into routine and bureaucracy, therefore lacking resources when new strategic advantages need to be discovered.

In the structuring proposed by Adizes, a manager must often abandon production tasks, which require a lot of time or technical know-how, and trust it to others who can see them through. Above all, they must focus on their work as an entrepreneur, facilitator and organizer.

When local elected officials concentrate on the general animation and coordination function, it falls to them to hire venturesome creatives to ignite innovation (what Jean-Marc Ayrault did in Nantes, or Martine Aubry in Lille, when that city became "European capital of culture" in 2004)

4.2.4.3. *Learning and innovation needs*

Whether we wish to reorient the tourism system or implement projects, understanding and accepting the necessary means (technologies, management methods) requires learning processes that should not be neglected.

Action tools or management schemes (Vaujany et al. 2016) must not be transmitted through coercion or imposition. No appropriation is possible using authoritarian methods. It is better to disseminate the necessary instrumentation by means of joint regulation (Reynaud 1989, 1994) by associating formal contributions and assimilations discussed in situ, gradually, with broad possibilities of adaptation to the work and action contexts of potential users.

Any contribution of means is to be discussed. It must be experienced as an opportunity rather than as a deprivation of autonomy. If the choice of tools is imposed, or if we incorporate tools that have barely been defined and are currently being tested[47], it is even worse. However, technical resources exist for making concerted choices, while respecting the specificities of the territories and the wishes of local actors.

Skill needs can be met through targeted training efforts. Tourism business managers have to understand that continuous learning is an imperative rule today, if only to adapt to mutations in their professions. For that matter, in no way would a car dealer (also a craftsman) ignore the transformations of the vehicles they repair. In tourism, things are evolving, even though not always spectacularly, and the slightest technical variation sometimes has significant consequences. In the digital sphere, while we have mainly taken into account the obligations in marketing, management control or the digital enrichment of activities, we have rarely assessed all the potential benefits. For example, how can we tempt customers to visit a place after watching an attractive video? How can we cleverly combine the face-to-face supply and the contributions of the sites offered remotely? Often, it is the influencers who are in charge of this job, but by doing so, they strip tourism professionals of the directions they wish to give to their teasing.

In traditional regions, or those without major tourist destinations, local energy needs to be reactivated. It is often appropriate to inject new blood, even accepting that there can be initial hostilities (as it happened with the neo-rurals in the 1970s). Had there not been a host of English "invaders" to take back agricultural land in Aquitaine or Gers, certain territories would be much less lively and certain festivals would never have seen the light of day. Newcomers must be able to settle in with their renegade ideas[48], and bring new solutions to those who, already established, can see the benefit in them. Sometimes these ideas are rejected; sometimes they are accepted.

This situation resembles that of active minorities (Couzineau-Zegwaard et al. (2022), who experiment with new products, practices or forms of management in their sector of activity. These groups have ways of thinking or acting that may seem out of sync with local habits. However, little by little, they start to take root and become sources of change and innovation. To do this, minorities must not make their relative deviance a source of conflict. As Moscovici indicated (1976), they must put in evidence the lasting benefits they derive from their action, while

47 In the digital world, the sale of software whose finishing touches or adaptation are not yet fully conclusive is quite frequent. This is done to rush the sale and get ahead of the competition (we have heard it from sales engineers several times at insider meetings).

48 According to the words of Mr. Huynh, a stylist who has recently restored a castle in Bordeaux: "There are little things in the spirit of the times one has to know how to capture" (*Le Monde Magazine*, 9 November, 2022).

appearing as autonomous (absence of sectarianism) and never criticizing those who act differently[49].

In a broader sense, innovation may be introduced by: astute entrepreneurs[50]; cultural cuttings installed in suitable places (the Louvre-Lens, Beaubourg in Metz, the new museums built in Lyon or Marseille); collective creativity encouraged by networks of actors around common sites, third places, living labs; inspiring elected officials who have both vision and strength for action, because they see broadly and have relatively free rein.

If it develops collective intelligence, any territory can become a learner, including its tourist collective. This intelligence is nourished by two types of resources:

– Storing available data, acquired knowledge we can still benefit from today. This makes it possible to chart a pathway, to move forward more quickly, to avoid previous mistakes and transform reactive behavior into proactive behavior. However, it is complicated. We often have to solve priming problems and the effective sharing of the information we wish to collect, valorize and operationalize.

– We activate the processes that will stimulate interactivity in networks. Relational intelligence contributes to the operationalization of project-oriented practical intelligence.

Tourists also participate in innovation by expressing their expectations. It is appropriate to train them on how to make choices. While they may have good reasons to be drawn to certain destinations, they can do so based on prejudices or on partial information. By sticking to narrow perspectives, they can make mistakes regarding the tourist objects that will best meet their needs.

Most innovations come from transfers of ideas from one sector to another. For example, biomimetics can lead to naturally air-conditioned hotels, or transition managers can be appointed in charge of a hotel-restaurant when the owner, or manager, is busy (illness, vacation, professional issues behind the need to take

---

49 This is how certain product innovations became widespread (pizzas in France in the 1970s, and more recently, the sales of kebabs in poor neighborhoods, for the price and friendliness), or Japanese restaurants in more affluent neighborhoods (for the exoticism of raw fish and the art of sushi); attractions with a slow start, then growing exponentially, as in escape games; visits to small cheese producers, following the reconversion of city dwellers to rural areas after 1968, etc.

50 On these aspects, we can also consult the journal *Management & Avenir* on "responsible tourism, a vector for environmental, social, economic and territorial innovation", coordination E. Leroux (2015, vol. 2, No. 76).

leave). It is better than innovating through "more of the same thing" (by offering even larger and softer beds, room vehicles for transportation from one hotel to another, or more greenery in common interiors[51]).

We must never consider that innovation is for the others[52]. Evidently, we can expect ideas from the outside, but it is better to also reflect on this with local actors. It is from within that we must "manage with audacity and creativity [...] to make things happen"[53], "rely on networks", "always reinvent oneself without ever knowing beforehand whether the audience will appreciate it" (taking risks). Also, for the employees themselves, we must "get them out of the straitjacket they are confined to for the sake of productivity".

Outside of the territories themselves, in a more open manner and less obsessed with short-term results, tourism can set up "colleges of the unknown", in the sense of Le Masson and Weil (2014, 2020). It all comes down to bringing together professionals and experts to explore possible futures, especially those that seem uncertain. This indulges us with time and broad-mindedness. We can explore every possibility by inviting specialists to discuss with them. We must avoid having appropriative aims, for the sake of moving quickly or benefiting unilaterally, when we ourselves are dependent on a company or a territory, and therefore subject to competitive games[54].

4.2.4.4. *The transition towards sustainable tourism*

Given the global context and polarization on climate change, the respect for ecosystems or decarbonization[55], it is urgent to embrace sustainable tourism as much as possible. This affects both the populations welcoming tourists in their territory and the measures to be taken in terms of energy, care given to sites and diversity. Also, this also impacts tourist behavior, which we will undoubtedly need to help transform.

---

51 This is what emerged from a large meeting of executives from global hotel chains in early 2023.

52 In May 2023, the city of Saumur put up for sale a former industrial building it wished to see transformed into a tourism venture. In the published advertisement, it "hoped to receive innovative offers". Now the question arises: could nothing emerge from internal reflection?

53 Lamarre, D. (spring 2022). "Une vie à élargir les horizons", CEO of various companies, including *Cirque du Soleil, interview* in the magazine *Gestion*, 47(1), 40–48.

54 The TIL could still have its place in this type of project. Bringing together two complementary possibilities: a "college of the unknown" to test long-term ideas, without any immediate constraint, and a "problem resolution" service, for sectors in need of innovation.

55 The IPCC report (2023) had never been so alarmist before. It speaks of the urgency of sobriety in all human and social conduct.

The balance between the needs of residents and those of tourists is sometimes resolved by a separation between living spaces and tourist sites (as is the case for Amnéville or Center Parcs). This is obviously not possible in territories where tourist activity interacts with the local population, whether or not they use it for their jobs. This happens in large cities, and when tourism is integrated or diffuse.

We have to make populations live together, avoiding the anger triggered by the abuse of excessive numbers, pollution and incivility generated by too large masses of visitors at the same time.

Sometimes, "overtourism[56]" pushes local systems and elected officials to react with protection demarketing, by discouraging all or part of the visitors, or by making them change dates, seasons, etc. There are all kinds of possibilities: reducing the visitor number (with a more qualitative supply) or offering a better distribution in space and time (through visit quotas, differentiated supply, an extension of places to be visited, better scheduled events)[57]. A vast number of tourists would agree to travel during quieter times[58]. Some hotels reorganize their rooms and fit them into houses or "human-sized" dwellings, thereby avoiding the "mass effect" of standard accommodation.

We must not forget that (along with peasant agriculture) tourism is one of the means for energizing the territory and reducing areas of economic disaffection or depopulation.

At the same time, we must reach a balance between tourism needs and ecological concerns, in anticipation of shortages or pollution problems, or to avoid the waste of equipment and material resources. Adapting to climate change requires contributing to energy decarbonization, while respecting the biodiversity of ecosystems. We find the same perspectives as in agriculture, since there too we must take into account the

56 Jessica Gourdon, "Étretat suffoque, rongé par le surtourisme": "while the Norman village has always seen its attendance increase from Easter until the end of summer, since the end of the Covid-19 crisis, it has been facing daily waves of thousands of visitors". *Le Monde*, 9–11 April, 2023, p. 15.

57 In Étretat (refer to the article on footnote 56 of this chapter), the town hall oscillates between maintaining activity and imposing regulation. Even the pebbles on the beach are stolen. To deal with this, measures have been taken: the creation of a mixed union within the framework of the *Grands Sites de France* label, distance from car parks, revamping the access to the cliffs, as if we had to "earn" the visit by walking.

58 Several countries accept teleworking at vacation spots for short periods (for example, one week). There is no need to hide when we are traveling.

state of the soil and changes in the climate[59]. There are major problems related to water (on the one hand, the risks of drought and fire, and on the other hand, the threat of submersion of the coasts). Tourism on the coast or in the mountains (winter sports) is greatly impacted. The same happens with those linked to food products (vineyards, various terroirs). In addition, we will have to integrate new transportation regulations.

In the face of these changes, tourists must also be taught to change their perspective and intentions. Tourism opens to desirable recreation. However, should we exacerbate commercial desire or social comparison games?[60] This question is obviously delicate, due to general profitability concerns and dependence on the intentions of tourists at the moment of choosing their destinations.

4.3. General conclusion: taking into account the spirit of the place!

Both produced by tourists and those who welcome them, tourist areas need lively and creative forces to garden their space and make it attractive. If there is not enough innovation, we must add more to the resources and, if possible, bring in new actors who can play the role of tourism or cultural entrepreneurs (as Nantes was able to do, or as we have seen in the cases where actors with special training and management power arrived).

By contributing to territorial coherence, tourism systems serve their clients, the professionals who receive them and the population as a whole. One of the interests of tourism is to provide a living for many persons at the site, without making jobs dependent on companies susceptible to poor remote management or strategic withdrawal for no local reason.

As in agriculture, we must limit the risk of "economic concentrations" with eccentric power, because this leads to partially losing the challenge of inhabiting a territory and bringing life to it. In the cases we analyzed, we identified this trend in relation to casinos, zoos, theme parks, hotels, restaurant chains and campsites. It is therefore important to attach them to the local site by using appropriate measures. We must also prevent tourism from being trapped, even when it is locally managed,

59 There are numerous articles and books on these questions, especially for agriculture, where the reflection is more structured than in tourism. Even for the Olympic Games, which aim to minimize their carbon footprint and have hired a "Director of Environmental Excellence", the promises are not easy to meet (see *Le Monde*, 25 May, 2023, "The difficult challenge of green Olympics", S. Mandard, pp. 20–21).

60 This is what the Icelandic poet Einar Mar Gudmundsson has denounced, with humor: "Then the mountain comes up and says: I am the mountain, but do not speak my name, do not say who I am. I would risk attracting tourists, all kinds of people who would run up my slopes, only to say that they climbed me".

by international platforms that take charge of bookings, financial management or customer evaluation. It can be alienating and enslaving. It is advisable to think about the best interests of each actor.

By reinventing autonomous proximities and protecting their originality, territories could differentiate their businesses and attractions, so that these are not repeated (from Saint-Jean-de-Monts to La Rochelle, from Périgueux to Nantes, from Amnéville to Lens, from Lake Geneva to Morbihan) in the form of the same brands, the same marketing styles, the same duplicated systems, available everywhere, and the same franchises installed in falsely differentiated urban planning. We always benefit from opening up to more variety, provided that it develops by relying on collective partnership.

To finish, let us take two types of examples of ways in which to revitalize a tourism system or amplify its creativity.

4.3.1. *Example of Amnéville*

How to regenerate Amnéville? The site had grown into a conglomerate with a collective interest because it made it possible to establish a growth dynamic. However, from cyclical uncertainties to institutional blunders, it was turned into an agglomeration of opportunities; certain actors benefited from local infrastructure (SDPL, etc.), while operating centrifugally or separately.

So, relying on public capital to strengthen certain facilities, develop infrastructure and build overall communication, how can we shape the project while rediscovering a dynamic of shared interests?

The best way is to proceed as in collective construction "sites", where the actors are obligatory partners, although institutionally belonging to external companies with their own strategy, because they share proximity challenges.

Following an exchange with a person responsible for operational reflection on the site's future, we could identify six vectors of strategic solidarity (Box 4.3).

– Overall communication (slogans and basic storytelling, campaigns and promotional materials).

– Management of common infrastructure (roads, signs, basic equipment).

– Management of events and entertainment on the site (integrating seasonal aspects, certain festive moments).

– Management of the interstitial fabric, for example in the creation of integrated accommodation (housing, bungalows, etc.).

– Internal synergistic effects (thanks to the specific quality of the place).

– The broader synergistic effects (the department and its need to host major sporting events, the region and the business networks engaged by Amnéville).

Box 4.3. *Vectors*

4.3.2. *Productive effects of a gradual diffusion of tourism initiatives*

Another direction is to exploit the destabilizing impact of an active subset to encourage local competitors to take initiatives. We see it in the diffuse system *Autour du Louvre Lens*. Local tensions and jealousies are used for instigating reactive creativity (healthy emulation, from which the tourists or the local residents may benefit). Valenciennes and Douai revived their museums and developed events within the same territory. On the other hand, the restoration of the mining districts (as part of the UNESCO World Heritage List) has incorporated cultural buildings as support for exhibitions or festivals.

The Nantes extensions (as part of the *Voyage à Nantes*) have boosted the Loire estuary. Dedicated to shipbuilding and aeronautics, the town of Saint-Nazaire has developed its seaside, multiplying cultural events. A former sub-prefecture and outport of Nantes – and after being asleep for a century – Paimboeuf emerged from its slumber when it was attached to the estuary's route and the bridge over the Loire reactivated it. New forms of mobility were used. North of La Baule, Batz and Le Croisic benefited from the TGV, while La Turballe doubled its anchorage port. This whole region benefited from the Parisian escapes during the Covid-19 crisis, and according to certain predictions, it is going to be a region relatively protected from climatic risks.

All this brings to mind what mathematics has described as transformative percolation[61].

To this, let us add two cases that were not studied in Chapter 1, but which are quite illustrative of construction by means of regenerative islets.

In the Perpignan region, where many vineyards are in danger of drought, small farms are adapting their wines to people's tastes and organic production, while testing methods to defy the climate that is threatening their terroir.

61 The act of changing structure from separate elements that end up relating to one another, for example, by passing from a diffuse disorder to a network and thereby acquiring more systemic properties.

In Banyuls, the "9 Caves" are a traditional site, turned into cellars by a winemaker from Loir-et-Cher, who convinced the locals and a Japanese couple. He allied himself with the goodwill of local residents and foreigners lured by the region (former tourists transformed into resource providers). The place includes two Dutch restaurateurs who had an excellent restaurant in Amsterdam, but desired to live in Roussillon.

We find this logic for the dozen taverns around Angers (e.g. in La Daguenière, Sainte-Gemmes, Bouchemaine-La Pointe, Cantenay-Épinard, Briollay, and Seiches). Listed as such on the Internet, they are always full, some for the hot season, others all year round, with real culinary milestones, a social atmosphere and a production of shows; whereas, the restaurateurs down town, along the banks of the Maine[62] (despite receiving aid from the town hall), content themselves with supplying heated dishes to their customers, believing that the charm of the place will suffice in itself.

According to Albert Hirschman[63]: "change is not something that will happen with certainty, it is an opportunity to be seized". It is "a passion for the possible". There is no general determinism (but trends or opportunities). There is no one best way (Taylorian notion, if ever there was one). Change does not go without useful paradoxes; for example, it is sometimes useful to "put the cart before the horse" (taking advantage of sequences which in theory would not have had their place at that moment in an evolutionary process); or what we thought were "obstacles to development" can be seized as potential levers for evolution (as blessings in disguise). Finally, "change requires uncertainties and surprises", because if it is implemented within a rationality that is too formal, too predictable, it has every chance of being blocked or "suppressed in advance by established interests".

62 This is the Slipper's Dock (*Cale de la Savate*), opposite the castle. The town hall brought together traders with two goals in mind: expansion and an aesthetic harmonization of the terraces.

63 Hirschman, A. (author of *Exit, Voice and Loyalty*) (1994). *L'analyse du changement*. In *L'analyse stratégique. Sa genèse, ses applications et ses problèmes actuels. Autour de Michel Crozier*, Pavé, F. (ed.). Le Seuil, Paris, pp. 15–20.

References

Adizes, I. (2004). *The Ideal Executive: Why You Cannot Be One and What To Do About It.* Adizes Institute, Carpinteria.

Allard Poesi, F. and Maréchal, G. (2014). Construction de l'objet de recherche. In *Méthodes de recherche en management*, Thiétart, R.-A. (ed.). Dunod, Paris.

Allison, G.T. (1971). *The Essence of Decision.* Little, Brown and Company, Boston.

Ancelin-Schützenberger, A. (1971). *Vocabulaire de psychosociologie de groupe et de formation.* EPI, Paris.

Ansell, C. and Gash, A. (2007). Collaborative governance in theory and practice. *Journal of Public Administration Research and Theory*, 8, 543–571.

Audet, M. and Malouin, J.-L. (1986). *La production des connaissances scientifiques de l'administration.* Presses de l'Université Laval, Quebec.

Augé, M. (1997). *L'impossible voyage. Le tourisme et ses images.* Éditions Payot et Rivages Poche Petite Bibliothèque, Paris.

Avenier, M. and Schmitt, C. (2007). Élaborer des savoirs actionnables et les communiquer à des managers. *Revue française de gestion*, 174, 25–42. doi: 10.3166/rfg.174.25-42.

Ballina, F.J. (2021). Conceptualization of tourist destinations using the "SmartMax" approach: An application to the Gijon destination (Spain). *Journal of Destination Marketing & Management*, 19. doi: 10.1016/j.jdmm.2020.100459.

Barney, J.B. (1991). Firm resources and sustained competitive advantage. *Journal of Management*, 17(1), 99–120.

Barney, J.B., Wright, M., Ketcher, D. (2001). Resource-based theories of competitive advantage: A ten-year retrospective on the resource-based view. *Journal of Management*, 27(6), 625–641.

Barouch, G. (1994). La rhétorique des monographies. In *L'analyse stratégique*, Pavé, F. (ed.). Le Seuil, Paris.

Barouch, G. (2001). Médiateur au service des organisations. *Journal de l'école de Paris*, 30.

Barthel, P.A. (2009). Faire la preuve de l'urbanisme durable : les enjeux de la régénération de l'île de Nantes. *Vertigo*, 9(2). doi: 10.4000/vertigo.8699.

Barthélémy, M. (2022). *Le monde des réseaux.* Odile Jacob, Paris.

Bartoli, A. (2012). Préface. In *La crise du management public, comment conduire le changement ?*, Trosa, S. (ed.). De Boeck, Paris.

Bédé, S. (2015). Les contrats de destination : nouveau souffle ou nouvelle ère de la gouvernance des destinations ? *Management & avenir*, 77, 119–136.

Bernard, G. (1996). La crise du cognac dans les années 1990. *Norois*, 172, 767–774.

Bertho-Lavenir, C. (2007). La visite du château. In *Made in Chambord*, Perlès, V. (ed.). Éditions du Patrimoine, Paris.

Blake, R. and Mouton, J. (1980). *Les deux dimensions du management.* Éditions d'Organisation, Paris.

Bonneau, M. (1973). Loisirs et fêtes dans le Saumurois : méthodologie et résultats. Working document, Travaux de l'Institut de Géographie de Reims, Reims, 13–14, 65–75.

Bonneau, M. (1975). Les problèmes du développement touristique dans une commune rurale : le cas de Montsoreau, station ligérienne. *Hommes et terres du nord*, 1, 3–28.

Botti, L., Clergeau, C., Peypoch N. (2019). La destination touristique. In *La recherche en management du tourisme*, Clergeau, C. and Peypoch, N. (eds). Vuibert/FNEGE/ Recherche, Paris.

Bouba-Olga, O. and Grossetti, M. (2018). Réflexion sur les dimensions géographiques et temporelles des processus territoriaux. *Revue d'économie régionale & urbaine*, 5(6), 1393–1406.

Bouba-Olga, O., Guimond, B., Perrier, B. (2010). Cognac, analyse d'une grappe d'activité. Study report for the Direction Régionale de l'Equipement Poitou-Charentes (DRE) [Online]. Available at: http://crief.labo.univ-poitiers.fr/IMG/pdf/Cognac_analyse_d_une_grappe_d_activites.pdf.

Bouba-Olga, O., Ferru, M., Guimond, B. (2012). Organisation des activités et dynamiques territoriales : éléments d'analyse et application aux bassins de Cognac et de Châtellerault. *Revue d'économie régionale et urbaine*, 2, 173–191.

Boudinet, J.P. (1996). *Anthropologie du projet.* PUF, Paris.

Boudon, R. (2003). *Raison, bonnes raisons.* PUF, Paris.

Boyer, M. (2005). *Histoire générale du tourisme du XVI*e *au XXI*e *siècle.* L'Harmattan, Paris.

Brandenburger, A. and Nalebuff, B.J. (1995). *The Right Game: Use Game Theory to Shape Strategy.* Harvard Business Review Press, Brighton.

Bréchet, J.P. and Desreumaux, A. (2018). *Repenser l'entreprise. Une théorie de l'entreprise fondée sur le projet.* Presses Universitaires du Septentrion, Lille.

Bréchet, J.-P., Charreaux, G., Montmorillon, B., Desreumaux, A. (2015). L'entreprise, son projet, sa gouvernance : éléments d'une vision partenariale. *Économies et sociétés*, 49(1), 33–65.

de Bruyn, C. and Fernández, A. (2012). Tourism destination governance. In *Knowledge Management in Tourism: Policy and Governance Applications*, volume 4, Fayos-solà, E., Albino Matos de Silva, J., Jafari, J. (eds). Emerald Publishing Limited, Bingley.

Butler, W. (1980). The concept of a tourist area cycle of evolution: Implications for management of resources. *Canadian Geographer*, 24(1), 5–12.

Cassou-Mounat, M. (1977). La vie humaine sur le littoral des Landes de Gascogne. Thesis, Université de Bordeaux 2, Bordeaux.

Cautrès, B. (2023). Culture du clash et manque de civilité ont des effets négatifs sur le crédit accordé au politique. *Le Monde* [Online]. Available at: https://www.lemonde.fr/idees/article/2023/03/14/bruno-cautres-politiste-culture-du-clash-et-manque-de-civilite-ont-des-effets-negatifs-sur-le-credit-accorde-au-politique_6165391_3232.html.

Cazes, G. and Potier, F. (1996). *Le tourisme urbain*. Presses Universitaires de France, Paris.

Clary, D. (1993). *Le tourisme dans l'espace français*. Masson, Paris.

Clerc, C. (2020). Structure et diversité des modèles actuels de gouvernement d'entreprise. Report for OIT [Online]. Available at: https://www.ilo.org/wcmsp5/groups/public/---europe/---ro-geneva/---ilo-paris/documents/publication/wcms_758031.pdf.

Clergeau, C. and Violier, P. (2012). Le concept de cluster est-il soluble dans le tourisme ? *Téoros*, 31(2), 60–71.

Clergeau, C., Pickel-Chevalier, S., Violier, P., Grefe, G. (2015). Naissance d'un cluster touristique équin ? L'exemple de Saumur (France). *Mondes du tourisme*, 188–204.

Colin, J. and Meschi, P.-X. (1998). Mobilisation de ressources et inertie structurelle, le cas de la Royale. In *Décisions et gestion*, Spiteri, P. (ed.). Presses de l'Université de Sciences Sociales de Toulouse, Toulouse.

Coussié, J.-V. (2011). *Le cognac, un produit régional, un marché mondial*. Self-published.

Couzineau-Zegwaard, E., Meier, O., Tarnaud, L. (2022). *Minorités en mode majeur*. Éditions EMS, Caen.

Crozier, M. (1979). *On ne change pas la société par décret*. Grasset, Paris.

Crozier, M. (1994). Le problème de l'action collective. L'éducation du regard. In *L'Analyse stratégique*, Pavé, F. (ed.). Le Seuil, Paris.

Crozier, M. and Friedberg, E. (1977). *L'acteur et le système*. Le Seuil, Paris.

Cusin, J. and Loubaresse, E. (2015). L'interclustering, de la communauté de pratique aux réseaux d'innovation. *Revue française de gestion*, 246(1), 13–39.

Darbellay, F., Clivaz, C., Nahrath, S., Stock, M. (2011). Approche interdisciplinaire du développement des stations touristiques. Le capital touristique comme concept opératoire. *Mondes du tourisme*, 4 [Online]. Available at: http://journals.openedition.org/tourisme/543.

Daubet, M. (1974). Les structures d'hébergement touristique en Haut-Quercy. *Revue géographique des Pyrénées et du Sud-Ouest*, 45(3), 297–316.

David, P.A. (1985). Clio and the economics of Qwerty. *The American Economic Review*, 75(2), 332–337.

Delbru, R. and Genty, M. (1986). Regards sur le tourisme en Périgord au milieu des années 1980. *Revue géographique des Pyrénées et du Sud-Ouest*, 57(4), 541–558.

Deprest, F. (1997). *Enquête sur le tourisme de masse : l'écologie face au territoire.* Éditions Belin/Mappemonde, Paris.

Desreumaux, A. (1992). *Structures d'entreprise.* Vuibert, Paris.

Dion, R. (1978). *Le Val de Loire, étude de géographie régionale.* Laffitte, Marseille.

Duhamel, P. (2003). Lieux. In *Le tourisme. Acteurs, lieux et enjeux*, Stock, M. (ed.). Éditions Belin Sup Géographie, Paris.

Duhamel, P. (2018). *Géographie du tourisme et des loisirs. Dynamiques, acteurs, territoires.* Armand Colin, Paris.

Duhamel, P. and Violier, P. (2009). *Tourisme et littoral. Un enjeu du monde.* Éditions Belin Sup-Tourisme, Paris.

Dumont, G.F. (2012). *Diagnostic et gouvernance des territoires. Concepts, méthode, application.* Armand Colin, Paris.

Durand, G. and Fererol, M.E. (2009). Le tourisme, positionnement stratégique au sein d'une métropole : le choix d'Evian vis-à-vis de la métropôle transfrontalière franco-valdo-genevoise. *L'Information géographique*, 3, 23–45.

Elias, N. and Dunning, E. (1995). *Sport et civilisation. La violence maîtrisée.* Fayard, Paris.

Everaere, C. (2007). *Autonomie et collectifs de travail.* ANACT, Paris.

Frederick, E.E. and Trist, E.L. (2017). Des systèmes sociotechniques à l'écologie sociale des organisations. In *Les grands auteurs en management*, Charreire Petit, S. and Huault, I. (eds). Editions EMS, Caen.

Friedberg, E. (1994). Le raisonnement stratégique comme méthode d'analyse et comme outil d'intervention. In *L'analyse stratégique*, Pavé, F. (ed.). Le Seuil, Paris.

Frochot, I. and Legohérel, P. (2018). *Marketing du tourisme. Construire une stratégie efficace.* Dunod, Paris.

Gaffard, J.-L. and Romani, P.M. (1990). À propos de la localisation des activités industrielles : le district marshallien. *Revue française d'économie*, 5(3), 171–185.

Girard, E. (2016). Le cognac : entre identité nationale et produit de la mondialisation. *Cybergeo* [Online]. Available at: http://journals.opendedition.org/cybergeo/27595.4.

Glon, B. and Pecqueur, P. (2016). *Au cœur des territoires créatifs, proximité et ressources territoriales.* Presses Universitaires de Rennes, Rennes.

Grafov, N. (2018). Le Louvre-Lens. Master's Thesis, Université de Lille, Lille.

Granovetter, M. (2000). The strength of weak ties. *American Journal of Sociology*, 1973(78).

Grosseti, M. (2004). *Sociologie de l'imprévisible. Dynamiques de l'activité et des formes sociales.* Presses Universitaires de France, Paris.

Guérout, J. and Mauduit, X. (eds) (2023). *Histoire des préjugés.* Les Arènes, Paris.

Guibert, C. (2016). Les déterminants dispositionnels du "touriste pluriel". Expériences, socialisations et contextes. *Sociologies, Theory and Research* [Online]. Available at: http://sociologies.revues.org/5688.

Hatchuel, A. (1997). Comment penser l'action collective ? Théorie des mythes rationnels. In *L'action collective*, Tosel, A. and Damien, R. (eds). Presses Universitaires de Franche-Comté, Besançon.

Hatchuel, A. (2019). *Henri Fayol et la théorie du chef d'entreprise.* École des Mines, Paris.

Hatchuel, A. and Weil, B. (1992). *L'expert et le système. Gestion des savoirs et métamorphose des acteurs dans l'entreprise industrielle.* Economica, Paris.

Hazebroucq, J.-M. (2009). De l'espace touristique à la "destination touristique" : un "territoire" qui se projette. In *Destinations et territoires, Volume 1. Coprésence à l'œuvre*, Lemasson, J.P. and Violier, P. (eds). Presses de l'Université du Québec, Quebec.

Humair, C. (2007). Ville, tourisme et transport : la compagnie du chemin de fer Lausanne-Ouchy (1869–1914). *Entreprises et histoire*, 2(47), 11–25.

Joutard, P. (1986). *L'invention du Mont Blanc.* Gallimard, Paris.

Kadri, B., Khomsi, R.H., Bondarenko, M. (2011). Le concept de destination. Diversité sémantique et réalité opérationnelle. *Téoros*, 30(1), 12–24.

Kadri, B., Delaplace, M., Grenier, A.A., Roche, Y. (eds) (2022). *Vocabulaire du discours touristique.* Presses de l'Université du Québec, Quebec.

Knafou, R. (1992). L'invention du tourisme. In *Encyclopédie de la géographie*, Bailly, A., Ferras, R., Pumain, D. (eds). Economica, Paris.

Knafou, R. (1995). Incertitudes, paradoxes et ambiguïtés sur le tourisme diffus. In *Le tourisme diffuse*, Jamot, C. and Vitte, P. (eds). CERAMAC, Clermont-Ferrand.

Knafou, R. and Stock, M. (2013). Tourisme. In *Dictionnaire de la géographie et de l'espace des sociétés*, Lévy, J. and Lussault, M. (eds). Éditions Belin, Paris.

Lahire, B. (1998). *L'homme pluriel. Les ressorts de l'action.* Nathan, Paris.

Landier, H. (1991). *Vers l'entreprise intelligente.* Calmann-Lévy, Paris.

Laperche, B., Lima, M., Seulliet, É., Trousse, B. (2019). *Les écosystèmes d'innovation : regards croisés des acteurs clés*. L'Harmattan. Paris.

Lauriol, J. (1998). *Décision stratégique en action*. L'Harmattan. Paris.

Lazzarotti, O. (2010). Le tourisme, matière à penser de la science géographique. *Mondes du tourisme*, 1, 7–16. doi: 10.4000/tourisme.314.

Le Masson, P. and Weil, B. (2014). Réinventer l'entreprise : la gestion collégiale des inconnus communs non appropriables. In *L'entreprise : point aveugle du savoir*, Segrestin, B. (ed.). Éditions Sciences Humaines, Auxerre. doi: 10.3917/sh.segre.2014.01.0238.

Le Masson, P. and Weil, B. (2020). L'expertise de l'inconnu : les chemins d'une création non destructrice. *Entreprises et histoire*, 98, 5–14. doi: 10.3917/eh.098.0005.

Leloup, F. (2010). Le développement territorial et les systèmes complexes : proposition d'un cadre analytique. *Revue d'économie régionale et urbaine*, 687–705.

Lemoigne, J.-L. (1990). *La modélisation des systèmes complexes*. Dunod, Paris.

Lévy, A. (1965). *Psychologie sociale, textes fondamentaux*. Dunod, Paris.

L'Hostis, M. (2020). La diffusion du tourisme chinois en France : influence des représentations et du capital spatial. PhD Thesis, Université d'Angers, Angers.

Lipietz, A. (2001). *Pour le tiers secteur, l'économie sociale et solidaire. Pourquoi et comment*. La Découverte/La Documentation française, Paris.

Lorino, P. (2020). *Pragmatisme et étude des organisations*. Economica, Paris.

Louart, P. (1980). L'artisan producteur et sous-traitant. PhD Thesis, Université de Lille, Lille.

Lozato-Giotart, J.-P. (1987). *Géographie du tourisme, de l'espace regardé à l'espace consommé*. Masson, Paris.

Lussault, M. (2013). Lieu. In *Dictionnaire de la géographie et de l'espace des sociétés*, Lévy, J. and Lussault, M. (eds). Éditions Belin, Paris.

Martineau, R. (2017). What are management tools made of? The "listic" structure of managerial artifacts. *M@n@gement*, 20(3), 239–262. doi: 10.3917/mana.203.0239.

Martinet, A. (2016). Recherche en stratégie : un problème de granularité. *Revue française de gestion*, 256, 11–18. doi: 10.3166/rfg.2016.00047.

Maslow, A.H. (1963). *Vers une psychologie de l'être*. Fayard, Paris.

Mendez, A. (2010). *Processus, concepts et méthodes pour l'analyse temporelle en sciences sociales*. Bruylant-Academia, Louvain-la-Neuve.

Michel, B. (2020). Pour une ville culturelle et touristique : retour sur une stratégie nantaise. In *Atlas social de la métropole nantaise*, Madoré, F. and Rivière, J. (eds) [Online]. Available at: http://asmn.univ-nantes.fr/index.php?id=450.

Milon, R. (1960). Le tourisme dans le département du Lot. *Revue géographique des Pyrénées et du Sud-Ouest*, 31(4), 381–397.

Mintzberg, H. (1979). *The Structuring of Organizations*. Prentice Hall, Englewood Cliffs.

Mirloup, J. (1987). Les grandes zones de tourisme culturel : définition et problèmes de saturation. *Méditerranée*, 62(4), 5–10.

Mission Opérationnelle Transfrontalière (MOT) (2002). La place de la coopération transfrontalière dans le développement territorial communautaire. In *Assemblée génerale extraordinaire et cinquième assemblée génerale annuelle*. Comédie des Champs Elysées, 8 October, Paris.

MIT team (2002). *Tourisme 1. Lieux communs*. Éditions Belin, Paris.

MIT team (2005). *Tourisme 2. Moments de lieux*. Éditions Belin, Paris.

MIT team (2011). *Tourisme 3. La révolution durable*. Éditions Belin, Paris.

Moisy, L. (2001). L'espace de la ville ludique et touristique : approche à travers les pratiques spatiales des visiteurs (hébergements, itinéraires). *Géocarrefour*, 76(2), 107–113.

Mondou, V. and Violier, P. (2009). Projets, pratiques et lieux touristiques, quelles relations ? *Mappemonde*, 94(2) [Online]. Available at: http://mappemonde.mgm.fr/num22/articles/art09201/html.

Monod, J.-C. (2012). *Qu'est-ce qu'un chef en démocratie ?* Le Seuil, Paris.

Moore, J.-F. (1996). *The Death of Competition: Leadership and Strategy in the Age of Business Ecosystems*. Harper Business, New York.

Moreno, J.L. (1970). *Fondements de la sociométrie*. Paris, PUF.

Morice, J.R. and Violier, P. (2009). De l'événement culturel à la destination touristique : les cas de Lille et Nantes. *Bulletin de l'Association des géographes français*, 86(3), 377–387.

Morin, P. (1994). L'intervention dans l'entreprise et le développement des organisations. In *Méthodologie économique*, Mouchot, C. (ed.). Hachette, Paris.

Moscovici, S. (1976). *La psychanalyse, son image et son public*. PUF, Paris.

Muller, P. (2015). *La société de l'efficacité globale. Comment les sociétés modernes se pensent et agissent sur elles-mêmes*. PUF, Paris.

Nayar, V. (2010). *Employees First. Customers Second: Turning Conventional Management Upside Down*. Harvard Business Review Press, Brighton.

Nobile, D. (2021). Intégrer la complexité dans la question du développement territorial par une posture entrepreneuriale des élus : le cas de la SPL Destination Amnéville. *Gestion et management public*, 9(3), 9–27 [Online]. Available at: https://doi-org.buadistant.univ-angers.fr/10.3917/gmp.093.0009.

Nonaka, I. and Takeuchi, H. (1995). *The Knowledge-Creating Company: How Japanese Companies Foster Creativity and Innovation for Competitive Advantage*. Oxford University Press, Oxford.

Pagels, H. (1990). *Les rêves de la raison. L'ordinateur et les sciences de la complexité*. Dunod/InterEditions, Paris.

Penrose, E. (1959). *The Theory of the Growth of the Firm*. John Wiley & Sons, New York.

Pérez, R. (2007). Pierre Tabatoni et le management. *Revue française de gestion*, 180, 15–24.

Petit, J.-G. and Sanguin, A.-L. (2003). *Les fleuves de la France atlantique : identités, espaces, représentations, mémoires*. L'Harmattan, Paris.

Pickel Chevalier, S. (2014). *L'Occident face à la nature : à la confluence des sciences, de la philosophie et des arts*. Le Cavalier Bleu, Paris.

Pickel Chevalier, S. (2019). Tourism and equine heritage in France: The case study of the Cadre noir de Saumur and the Vendée Stud. *Current Issues in Tourism*, 24(1). doi: 10.1080/13683500.2019.1706459.

Pickel Chevalier, S. (2020). Mise en tourisme des Haras nationaux. De la perpétuation à la réinvention ? *Économie rurale, Société française d'économie rurale*, 374, 35–51. doi: 10.4000/economierurale.8247.

Pinte, J.-P. (2006). La veille informationnelle en éducation pour répondre au défi de la société de la connaissance au XXIe siècle : application à la conception d'une plateforme de veille et de partage de connaissance en éducation : commun@utice. PhD Thesis, Université de Marne-la-Vallée, Marne-la-Vallée.

Piriou, J. (2011). Destinations et positions de lieux de l'arc Lémanique. In *Tourisme et mondialisation*, Duhamel, P. and Kadri, B. (eds). Espaces, Paris.

Piriou, J. (2012). Enquête sur la région touristique : une recherche sur les pratiques spatiales de dimension régionale des acteurs du tourisme. PhD Thesis, Université d'Angers, Angers.

Piriou, J. (2016). La dimension régionale du système touristique : une lecture réticulaire par les pratiques spatiales des touristes au sein des "châteaux de la Loire". *Mondes du tourisme*. doi: 10.4000/tourisme.1276.

Piriou, J. (2018a). Partage de compétence de la promotion des destinations touristiques en France et en Suisse. Le cas de l'espace transfrontalier de l'Arc Lémanique. In *Le tourisme, un domaine de compétence partagée*, Fabry, N. and Spindler, J. (eds). L'Harmattan, Paris.

Piriou, J. (2018b). The Cognac vineyard area: Territorial quality as a focus for tourism development within the context of a globally consumed product. In *Gastronomy and Local Development: The Quality of Products, and Places Experiences*, Bellini, N., Clergeau, C., Etcheverria, O. (eds). Routledge, London.

Piriou, J. (2019). *The Tourist Region: A Co-Construction of Tourism Stakeholders*. ISTE Ltd, London, and John Wiley & Sons, New York.

Piriou, J. (2020a). Lieux en marge et mobilités touristiques. Différencier les lieux ordinaires des lieux à fonctions touristiques régionale. *Téoros* [Online]. Available at: http://journals.openedition.org/teoros/4477.

Piriou, J. (2020b). L'office de tourisme, une institution locale centenaire : la créativité pour adaptation à sa destination touristique. In *Organisations créatives et culturelles*, Baujard, C., Lagier, J., Montargot, N. (eds). ISTE Éditions, London.

Piriou, J., Ananian, P., Clergeau, C. (2017). *Tourisme et événementiel : enjeux territoriaux et stratégies d'acteurs*. Presses de l'Université du Québec, Quebec.

Pittion, J.-P. (2014). Saumur et les voyageurs britanniques au XVIIe siècle. *Annales de Bretagne et des Pays de l'Ouest*, 121(3), 97–117.

Porter, M.E. (1998). Clusters and the new economics if competition. *Harvard Business Review*, 76(6), 77–90.

Poschet, L., Wust, S., Bassand, M. (2001). La métropole lémanique. Final report, Pôle Rhône-Alpes-Suisse, Programme Interreg IIc, Institut de recherche sur l'environnement contruit, IREC, École Polytechnique fédérale de Lausanne, Lausanne.

Raffestin, C. (1980). *Pour une géographie du pouvoir*. ENS Éditions, Paris.

Rewerski, J. (1983). Les loisirs et monde souterrain en Anjou. *Norois*, 120, 565–570.

Reynaud, J.-D. (1989). *Les règles du jeu : l'action collective et la régulation sociale*. Dunod, Paris.

Reynaud, J.-D. (1994). L'acteur stratégique et sa légitimité. In *L'analyse stratégique. Sa genèse, ses applications et ses problèmes actuels. Autour de Michel Crozier*, Pavé, F., (ed.). Le Seuil, Paris.

Rinaldi, E. (2006). La construction des hôtels de Montreux et les italiens à la fin du XIXe siècle. *Revue historique vaudoise*, 114, 241–252.

Rokeach, M. (1973). *The Nature of Human Values*. The Free Press, New York.

Rousseau J.-J. (1953). *The Confessions*, Cohen J.M. (trans.). Penguin Books, London.

Saint-Arnaud, Y. (2009). *Les petits groupes, participation et communication*. Gaëtan Morin, Montreal.

Sceau, R. (1974). Evian-les-Bains, station thermale et touristique. *Revue de géographie de Lyon*, 49(1), 51–75.

Stock, M. (2008). Il mondo è mobile. In *L'invention du monde. Une géographie de la mondialisation*, Lévy, J. (ed.). Presses de Sciences Po, Paris.

Stock, M. (2017). *Les enjeux contemporains du tourisme. Une approche géographique*. Presses Universitaires de Rennes, Rennes.

Tabariés, M. (2005). Les apports du GREMI à l'analyse territoriale de l'innovation. *Cahiers de la Maison des sciences économiques*, 18.

Tabatoni, P. and Jarniou, P. (1975). *Les systèmes de gestion : politiques et structures*. Presses Universitaires de France, Paris.

Tanguy, C. and Uzunidis, D. (2016). Milieu innovateur et entrepreneuriat innovant : la force des proximités et des réseaux. *Technologie et innovation*, 16(1), 91–106.

Thiéry, S. and Sénard, J.D. (2023). Rapport des Assises du travail, 17 propositions pour reconsidérer le travail [Online]. Available at: https://travail-emploi.gouv.fr/IMG/pdf/assises_du_travail_-_rapport_des_garants.pdf.

Thorelli, H.B. (1986). *Networks: Between Markets and Hierarchies*. John Wiley & Sons, New York.

Tinsley, R. and Lynch, P. (2001). Small tourism business networks and destination development. *International Journal of Hospitality Management*, 20(4), 367–378.

Tissot, L. (2000). *Naissance d'une industrie touristique. Les Anglais et la Suisse au XIXe siècle*. Payot, Lausanne.

Tissot, L. (2007). Le tourisme : de l'utopie réalisée au cauchemar généralisé ? *Entreprises et Histoire*, 2(47), 5–10.

Torre, A. (2016). La figure du réseau : dimensions spatiales et organisationnelles. *Géographie, économie, société*, 18(4), 455–469.

Trosa, S. (2012). *La crise du management public. Comment conduire le changement ?* De Boeck, Paris.

Van de Ven, A.H. and Poole, M.S. (1995). Explaining development and change in organizations. *Academy of Management Review*, 20(3), 510–540.

Vaujany, F., Hussenot, A., Chanlat, J.F. (2016). *Théories des organisations, nouveaux tournants*. Economica, Paris.

Vincent, J. (2015). La Société d'aménagement de la Côte de Monts (SACOM) : du rêve "les pieds dans l'eau" au cauchemar urbain ? *Cahiers de l'histoire du pays maraîchin*, 1, 10–30.

Violier, P. (1998a). À la recherche du tourisme urbain : exploration nantaise. In *Le tourisme et la ville : expériences européennes*, Cazes, G. and Potier, F. (eds). L'Harmattan, Paris.

Violier, P. (1998b). Système des acteurs du tourisme et relations ville-campagne. *Norois*, 178, 325–336

Violier, P. (1999). Le Pays entre institution et développement local. *Norois*, 181, 153–161.

Violier, P. (2008). *Tourisme et développement local*. Éditions Belin Sup-Tourisme, Paris.

Violier, P. (2009). Proposition pour un modèle d'analyse de la dynamique des lieux touristiques. In *Destinations et territoires, Volume 1. Coprésence à l'œuvre*, Lemasson, J.-P. and Violier, P. (eds). Presses de l'Université du Québec, Quebec.

Violier, P. (2016). La troisième révolution touristique. *Mondes du tourisme*. doi: 10.4000/tourisme.1256.

Violier, P. (2017). Comment les individus habitent-ils touristiquement le Monde ? In *Les espaces du tourisme et des loisirs*, Fagnoni, E. (ed.). Armand Colin, Paris.

Violier, P. and Taunay, B. (2020). *The Tourist Places of the World*. ISTE Ltd, London, and John Wiley & Sons, New York.

Violier, P., Duhamel, P., Gay, J.C., Mondou, V. (2021). *Le tourisme en France 1 : approche globale*. ISTE Éditions, London.

Vlès, V. (1996). *Les stations touristiques*. Economica, Paris.

Worms, J.-P. (1994). Mais si, on peut changer la société par décret ! In *L'analyse stratégique*, Pavé, F. (ed.). Le Seuil, Paris.

Worms, J.-P. and Grémion, P. (1970). L'État et les collectivités locales. *Revue esprit*, 1, 20–35.

Index

Other titles from

iSTE

in

Science, Society and New Technologies

2024

BARTHES Angela, CIBIEN Catherine, ROMAGNY Bruno
Biosphere Reserves and Sustainable Development Goals 1: Scientific and Practical Educational Issues in the Mediterranean
(Territory Development Set – Volume 3)
Biosphere Reserves and Sustainable Development Goals 2: Issues, Tensions, Processes and Governance in the Mediterranean
(Territory Development Set – Volume 4)

KOHLMANN Émilie
Linking with Nature in the Digital Age
(Communication, Environment, Science and Society Set – Volume 1)

2023

APCHAIN Thomas
Favela Tours: Building Otherness in the Favelas of Rio de Janeiro
(Tourism and Mobility Systems Set – Volume 6)

CARDON Alain, EL HAMI Abdelkhalak
Fundamental Generation Systems: Computer Science and Artificial Consciousness, the Informational Field of Generation of the Universe, the Sixth Sense of Living Beings
(Digital Sciences Set – Volume 4)

CURULLI Irene G., IKIZ KAYA Deniz, KHAEFI Arghavan
Heritage Education for Climate Action
(Research in Architectural Education Set – Volume 2)

ELAMÉ Esoh
The Sustainable City in Africa Facing the Challenge of Liquid Sanitation
(Territory Development Set – Volume 2)

JURCZENKO Emmanuel
Climate Investing: New Strategies and Implementation Challenges

LYUBAREVA Inna, WALDECK Roger
Diversity of Methodological Approaches in Social Sciences: Example of the Analysis of Media and Online Information
(Modeling Methodologies in Social Sciences Set – Volume 3)

PARET Dominique, CRÉGO Pierre, SOLÈRE Pauline
Smart Patches: Biosensors, Graphene and Intra-Body Communications

VENTRE Daniel, GUILLOT Philippe
Electronic Communication Interception Technologies and Issues of Power

VENTRE Daniel, LOISEAU Hugo
Cybercrime during the SARS-CoV-2 Pandemic (2019-2022): Evolutions, Adaptations, Consequences
(Cybersecurity Set – Volume 3)

2022

AIT HADDOU Hassan, TOUBANOS Dimitri, VILLIEN Philippe
Ecological Transition in Education and Research

CARDON Alain
Information Organization of The Universe and Living Things: Generation of Space, Quantum and Molecular Elements, Coactive Generation of Living Organisms and Multiagent Model
(Digital Sciences Set – Volume 3)

CAULI Marie, FAVIER Laurence, JEANNAS Jean-Yves
Digital Dictionary

DAVERNE-BAILLY Carole, WITTORSKI Richard
Research Methodology in Education and Training: Postures, Practices and Forms
(Education Set – Volume 12)

ELAMÉ Esoh
Sustainable Intercultural Urbanism at the Service of the African City of Tomorrow
(Territory Development Set – Volume 1)

FLEURET Sébastien
A Back and Forth Between Tourism and Health: From Medical Tourism to Global Health
(Tourism and Mobility Systems Set – Volume 5)

KAMPELIS Nikos, KOLOKOTSA Denia
Smart Zero-energy Buildings and Communities for Smart Grids
(Engineering, Energy and Architecture Set – Volume 9)

2021

BARDIOT Clarisse
Performing Arts and Digital Humanities: From Traces to Data
(Traces Set – Volume 5)

BENSRHAIR Abdelaziz, BAPIN Thierry
From AI to Autonomous and Connected Vehicles: Advanced Driver-Assistance Systems (ADAS)
(Digital Sciences Set – Volume 2)

BÉRANGER Jérôme
Societal Responsibility of Artificial Intelligence: Towards an Ethical and Eco-responsible AI
(Technological Prospects and Social Applications Set – Volume 4)

CORDELIER Benoit, GALIBERT Oliver
Digital Health Communications
(Technological Prospective and Social Applications Set – Volume 5)

DOUAY Nicolas, MINJA Michael
Urban Planning for Transitions

GALINON-MÉLÉNEC Béatrice
The Trace Odyssey 1: A Journey Beyond Appearances
(Traces Set – Volume 4)

HENRY Antoine
Platform and Collective Intelligence: Digital Ecosystem of Organizations

KOLOKOTSA Denia, KAMPELIS Nikos
Smart Buildings, Smart Communities and Demand Response
(Engineering, Energy and Architecture Set – Volume 8)

LE LAY Stéphane, SAVIGNAC Emmanuelle, LÉNEL Pierre, FRANCES Jean
The Gamification of Society
(Research, Innovative Theories and Methods in SSH Set – Volume 2)

RADI Bouchaïb, EL HAMI Abdelkhalak
Optimizations and Programming: Linear, Non-linear, Dynamic, Stochastic and Applications with Matlab
(Digital Sciences Set – Volume 1)

2020

ALAKTIF Jamila, CALLENS Stéphane
Migration and Climate Change: From the Emergence of Human Cultures to Contemporary Management in Organizations

BARNOUIN Jacques
The World's Construction Mechanism: Trajectories, Imbalances and the Future of Societies
(Interdisciplinarity Between Biological Sciences and Social Sciences Set – Volume 4)

ÇAĞLAR Nur, CURULLI Irene G., SIPAHIOĞLU Işıl Ruhi, MAVROMATIDIS Lazaros E.
Thresholds in Architectural Education
(Engineering, Energy and Architecture Set – Volume 7)

DUBOIS Michel J.F.
Humans in the Making: In the Beginning was Technique
(Social Interdisciplinarity Set – Volume 4)

ETCHEVERRIA Olivier
The Restaurant, A Geographical Approach: From Invention to Gourmet Tourist Destinations
(Tourism and Mobility Systems Set – Volume 3)

GREFE Gwenaëlle, PEYRAT-GUILLARD Dominique
Shapes of Tourism Employment: HRM in the Worlds of Hotels and Air Transport
(Tourism and Mobility Systems Set – Volume 4)

JEANNERET Yves
The Trace Factory
(Traces Set – Volume 3)

KATSAFADOS Petros, MAVROMATIDIS Elias, SPYROU Christos
Numerical Weather Prediction and Data Assimilation
(Engineering, Energy and Architecture Set – Volume 6)

MARTI Caroline
Cultural Mediations of Brands: Unadvertization and Quest for Authority
(Communication Approaches to Commercial Mediation Set – Volume 1)

MAVROMATIDIS Lazaros
Climatic Heterotopias as Spaces of Inclusion: Sew Up the Urban Fabric
(Research in Architectural Education Set – Volume 1)

MOURATIDOU Eleni
Re-presentation Policies of the Fashion Industry: Discourse, Apparatus and Power
(Communication Approaches to Commercial Mediation Set – Volume 2)

SCHMITT Daniel, THÉBAULT Marine, BURCZYKOWSKI Ludovic
Image Beyond the Screen: Projection Mapping

VIOLIER Philippe, with the collaboration of TAUNAY Benjamin
The Tourist Places of the World
(Tourism and Mobility Systems Set – Volume 2)

2019

BRIANÇON Muriel
The Meaning of Otherness in Education: Stakes, Forms, Process, Thoughts and Transfers
(Education Set – Volume 3)

DESCHAMPS Jacqueline
Mediation: A Concept for Information and Communication Sciences
(Concepts to Conceive 21st Century Society Set – Volume 1)

DOUSSET Laurent, PARK Sejin, GUILLE-ESCURET Georges
Kinship, Ecology and History: Renewal of Conjectures
(Interdisciplinarity Between Biological Sciences and Social Sciences Set – Volume 3)

DUPONT Olivier
Power: A Concept for Information and Communication Sciences
(Concepts to Conceive 21st Century Society Set – Volume 2)

FERRARATO Coline
Prospective Philosophy of Software: A Simondonian Study
(Technological Prospects and Social Applications Set – Volume 3)

GUAAYBESS Tourya
The Media in Arab Countries: From Development Theories to Cooperation Policies

HAGÈGE Hélène
Education for Responsibility
(Education Set – Volume 4)

LARDELLIER Pascal
The Ritual Institution of Society
(Traces Set – Volume 2)

LARROCHE Valérie
The Dispositif: A Concept for Information and Communication Sciences
(Concepts to Conceive 21st Century Society Set – Volume 3)

LATERRASSE Jean
Transport and Town Planning: The City in Search of Sustainable Development

LELEU-MERVIEL Sylvie, SCHMITT Daniel, USEILLE Philippe
From UXD to LivXD: Living eXperience Design

LENOIR Virgil Cristian
Ethically Structured Processes
(Innovation and Responsibility Set – Volume 4)

LOPEZ Fanny, PELLEGRINO Margot, COUTARD Olivier
Local Energy Autonomy: Spaces, Scales, Politics
(Urban Engineering Set – Volume 1)

METZGER Jean-Paul
Discourse: A Concept for Information and Communication Sciences
(Concepts to Conceive 21st Century Society Set – Volume 4)

MICHA Irini, VAIOU Dina
Alternative Takes to the City
(Engineering, Energy and Architecture Set – Volume 5)

PÉLISSIER Chrysta
Learner Support in Online Learning Environments

PIETTE Albert
Theoretical Anthropology or How to Observe a Human Being
(Research, Innovative Theories and Methods in SSH Set – Volume 1)

PIRIOU Jérôme
The Tourist Region: A Co-Construction of Tourism Stakeholders
(Tourism and Mobility Systems Set – Volume 1)

PUMAIN Denise
Geographical Modeling: Cities and Territories
(Modeling Methodologies in Social Sciences Set – Volume 2)

WALDECK Roger
Methods and Interdisciplinarity
(Modeling Methodologies in Social Sciences Set – Volume 1)

2018

BARTHES Angela, CHAMPOLLION Pierre, ALPE Yves
Evolutions of the Complex Relationship Between Education and Territories
(Education Set – Volume 1)

BÉRANGER Jérôme
The Algorithmic Code of Ethics: Ethics at the Bedside of the Digital Revolution
(Technological Prospects and Social Applications Set – Volume 2)

DUGUÉ Bernard
Time, Emergences and Communications
(Engineering, Energy and Architecture Set – Volume 4)

GEORGANTOPOULOU Christina G., GEORGANTOPOULOS George A.
Fluid Mechanics in Channel, Pipe and Aerodynamic Design Geometries 1
(Engineering, Energy and Architecture Set – Volume 2)
Fluid Mechanics in Channel, Pipe and Aerodynamic Design Geometries 2
(Engineering, Energy and Architecture Set – Volume 3)

GUILLE-ESCURET Georges
Social Structures and Natural Systems: Is a Scientific Assemblage Workable?
(Social Interdisciplinarity Set – Volume 2)

LARINI Michel, BARTHES Angela
Quantitative and Statistical Data in Education: From Data Collection to Data Processing
(Education Set – Volume 2)

LELEU-MERVIEL Sylvie
Informational Tracking
(Traces Set – Volume 1)

SALGUES Bruno
Society 5.0: Industry of the Future, Technologies, Methods and Tools
(Technological Prospects and Social Applications Set – Volume 1)

TRESTINI Marc
Modeling of Next Generation Digital Learning Environments: Complex Systems Theory

2017

ANICHINI Giulia, CARRARO Flavia, GESLIN Philippe, GUILLE-ESCURET Georges
Technicity vs Scientificity: Complementarities and Rivalries
(Interdisciplinarity Between Biological Sciences and Social Sciences Set – Volume 2)

DUGUÉ Bernard
Information and the World Stage: From Philosophy to Science, the World of Forms and Communications
(Engineering, Energy and Architecture Set – Volume 1)

GESLIN Philippe
Inside Anthropotechnology: User and Culture Centered Experience
(Social Interdisciplinarity Set – Volume 1)

GORIA Stéphane
Methods and Tools for Creative Competitive Intelligence

KEMBELLEC Gérald, BROUDOUX Evelyne
Reading and Writing Knowledge in Scientific Communities: Digital Humanities and Knowledge Construction

MAESSCHALCK Marc
Reflexive Governance for Research and Innovative Knowledge
(Responsible Research and Innovation Set - Volume 6)

PARK Sejin, GUILLE-ESCURET Georges
Sociobiology vs Socioecology: Consequences of an Unraveling Debate
(Interdisciplinarity Between Biological Sciences and Social Sciences Set – Volume 1)

PELLÉ Sophie
Business, Innovation and Responsibility
(Responsible Research and Innovation Set – Volume 7)

2016

ANDRÉ Michel, SAMARAS Zissis
Energy and Environment
(Research for Innovative Transports Set – Volume 1)

BLANQUART Corinne, CLAUSEN Uwe, JACOB Bernard
Towards Innovative Freight and Logistics
(Research for Innovative Transports Set – Volume 2)

COHEN Simon, YANNIS George
Traffic Management
(Research for Innovative Transports Set – Volume 3)

EL FALLAH SEGHROUCHNI Amal, ISHIKAWA Fuyuki, HÉRAULT Laurent, TOKUDA Hideyuki
Enablers for Smart Cities

GIANNI Robert
Responsibility and Freedom: The Ethical Realm of RRI
(Responsible Research and Innovation Set – Volume 2)

GRUNWALD Armin
The Hermeneutic Side of Responsible Research and Innovation
(Responsible Research and Innovation Set – Volume 5)

LENOIR Virgil Cristian
Ethical Efficiency: Responsibility and Contingency
(Responsible Research and Innovation Set – Volume 1)

PELLÉ Sophie, REBER Bernard
From Ethical Review to Responsible Research and Innovation
(Responsible Research and Innovation Set – Volume 3)

REBER Bernard
Precautionary Principle, Pluralism and Deliberation: Science and Ethics
(Responsible Research and Innovation Set – Volume 4)

TORRENTI Jean-Michel, LA TORRE Francesca
Materials and Infrastructures 1
(Research for Innovative Transports Set – Volume 5A
Materials and Infrastructures 2
(Research for Innovative Transports Set – Volume 5B)

VENTRE Daniel
Information Warfare – 2nd edition

YANNIS George, COHEN Simon
Traffic Safety
(Research for Innovative Transports Set – Volume 4)

2015

BRONNER Gérald
Belief and Misbelief Asymmetry on the Internet

2009

VENTRE Daniel
Information Warfare

Printed and bound by CPI Group (UK) Ltd, Croydon, CR0 4YY
16/09/2026
14963082-0006